CLIMATE JUSTICE

# CLIMATE JUSTICE

## WHAT RICH NATIONS OWE THE WORLD—AND THE FUTURE

CASS R. SUNSTEIN

The MIT Press
Cambridge, Massachusetts
London, England

The MIT Press would like to thank the anonymous peer reviewers who provided comments on drafts of this book. The generous work of academic experts is essential for establishing the authority and quality of our publications. We acknowledge with gratitude the contributions of these otherwise uncredited readers.

This book was set in Stone Serif and Stone Sans by Jen Jackowitz. Printed and bound in the United States of America.

Library of Congress Cataloging-in-Publication Data

Names: Sunstein, Cass R., author.
Title: Climate justice : what rich nations owe the world—and the future / Cass R. Sunstein.
Description: Cambridge, Massachusetts : The MIT Press, 2025. | Includes bibliographical references and index.
Identifiers: LCCN 2024019383 (print) | LCCN 2024019384 (ebook) | ISBN 9780262049467 (hardcover) | ISBN 9780262381505 (pdf) | ISBN 9780262381512 (epub)
Subjects: LCSH: Climate justice. | Environmental ethics.
Classification: LCC GE220 .S86 2025 (print) | LCC GE220 (ebook) | DDC 304.2/8—dc23/eng/20240808
LC record available at https://lccn.loc.gov/2024019383
LC ebook record available at https://lccn.loc.gov/2024019384

10 9 8 7 6 5 4 3 2 1

# CONTENTS

# INTRODUCTION

Everyone knows, or should know, that climate change is helping to create horrors: flooding, wildfire, extreme heat, drought, and much more. What such words do not adequately capture are the concrete harms: deaths; illnesses; losses of jobs, income, and opportunity; fear, stress, and sometimes terror. Pictures would do much better.

Figures 0.1, 0.2, 0.3, and 0.4 show a few.

**FIGURE 0.1**
BLM Oregon/Washington, *Wildfire in the Pacific Northwest.* Licensed under CC BY 2.0.

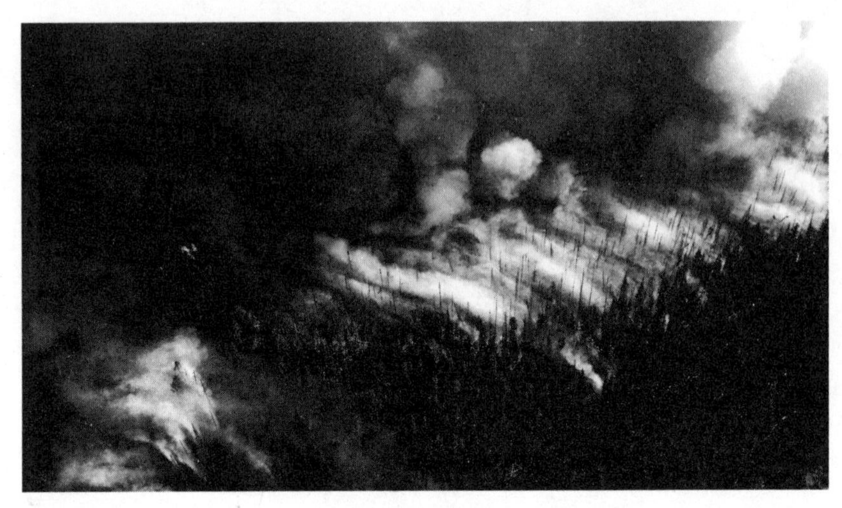

FIGURE 0.2
NPS Climate Change Response, *Wildfire*. Marked with Public Domain Mark 1.0.

FIGURE 0.3
U.S. Institute of Peace, *Flooding Challenges Pakistan's Government and the International Community*. Marked with CC BY 2.0.

While no one is immune from the risks of climate change, some people, and some nations, are far more vulnerable than others. People in Pakistan face greater risks than do people in Canada. People in Afghanistan are much more vulnerable than are people in Sweden. People in Chad are much more

FIGURE 0.4
Graeme Maclean, *Extreme Heat Warning Sign in Death Valley, California*. Marked with CC BY 2.0.

vulnerable than are people in Germany. People in Somalia, the Democratic Republic of Congo, Nigeria, and Ethiopia are much more vulnerable than are people in France, Denmark, and the United States.

It is also clear that some nations have contributed far more to the problem of climate change than others. The United States wins the prize for all-time greatest contributor. China is, by far, the greatest annual contributor. It follows that people in some countries will experience untold suffering and that people in other nations are largely responsible for that.

As of this writing, the world's top five emitters are the United States, China, Russia, Brazil, and India. Since 1990, their emissions have been estimated, on one account, to have produced a staggering $6 trillion in income losses, which is comparable to 11 percent of the annual global gross domestic product.[1] Let us pause over that estimate without necessarily crediting it. Whatever we think of any particular set of numbers, no one doubts that the largest emitters are mostly responsible for the losses. Importantly, the distribution of those losses is highly unequal. A disproportionate amount is borne by low-income, low-emitting countries. The high-emitting countries have gained a lot from their emitting activities and have lost relatively less.

How should we think about that?

Here is my starting point: each person should be counted equally, no matter *where* they live, and no matter *when* they live. As we shall see, that claim, appropriately qualified, has implications for an assortment of concrete policies related to climate change.

I take my inspiration here from John Stuart Mill, who emphasized that utilitarianism, his preferred approach to ethical questions, does not call for selfish behavior and indeed does not even authorize it. In Mill's account, the "utilitarian standard of what is right in conduct, is not the agent's own happiness, but that of all concerned."[2] Having said that, Mill started to soar: "As between his own happiness and that of others, utilitarianism requires him to be as strictly impartial as a disinterested and benevolent spectator. In the golden rule of Jesus of Nazareth, we read the complete spirit of the ethics of utility. To do as you would be done by, and to love your neighbour as yourself, constitute the ideal perfection of utilitarian morality."

In the context of climate change, it is probably too much to expect people to "love" their neighbors as themselves. But it might not be too much to ask people to aim to do as they would be done by. (I have not mentioned nonhuman animals, but they matter too. On that one, let's stomp our foot a bit. Harms to nonhuman animals must be counted in the overall assessment of the welfare effects of climate change, though I will not have much to say on that important topic here.)

There are cautionary notes, of course—some of them involving self-interest, some of them involving people's moral intuitions, some of them involving what is practical. Most important, perhaps, is *nationalism*. I love the United Kingdom, for example, and I am lucky enough to have an affiliation with one of its great universities, but the people of that admirable nation do not think that their responsibility to the people of Ethiopia is equivalent to their responsibility to the people of the United Kingdom. Nationalism raises a host of hard questions, and we shall explore some of them here. For the record: I do not oppose nationalism as such. Still, the ethics of nationalism is inconsistent with the spirit of the ethics of utility and with the golden rule of Jesus of Nazareth.

## SOME PERSONAL WORDS

This book is a product of personal experiences. The project was born, I am pretty sure, in February 2009. At the time, I was a senior adviser to the director of the Office of Management and Budget, having been nominated to serve as administrator of the Office of Information and Regulatory Affairs (and awaiting Senate confirmation, which did not come until September).

The Obama Administration was highly focused on climate change, and for many of us, the central questions were simple: What were we going to do about it? And when?

To make some progress on these questions, I had lunch at the White House dining room (the "Mess," as it is called) with Michael Greenstone, who was serving as chief economist in the Council of Economic Advisers. After a few pleasantries (we are friends), we discussed whether and how to regulate greenhouse gas emissions, and more specifically, how to monetize the benefits of such regulations. Should we regulate motor vehicles? Power plants? How much?

The discussion of monetization was essential. Here's the reason: *You can't decide how stringently to regulate greenhouse gas emissions unless you decide, if only implicitly, on the value of emissions reductions.* Suppose that you can eliminate two million tons of carbon dioxide emissions at a price. What price would be so high that you would decide not to do it? What price would be so low that it is obvious that you would do it? Are those political questions? What kinds of questions are they?

Greenstone and I decided that the United States government needed a *social cost of carbon* (SCC). The basic idea was hardly original. It had been elaborated in the economic literature and even explored in court; there was nothing new in the suggestion that carbon had a social cost. But as a nation, the United States government did not have, and had never had, a social cost of carbon. That did not make sense. Nor would it make sense if the Environmental Protection Agency had one, and the Department of Transportation had another one, and the Department of Agriculture had a third.

To come up with a social cost of carbon, public officials have to answer a host of questions. One of the most fundamental is whether and how to count losses to people outside of the United States. If emissions from the United States impose harms on France, Chad, Afghanistan, or Italy, should those harms count? Back in 2009, that question was closely connected with academic work that I had been doing on ethical issues central to climate change. These included issues of distributive justice: Should rich countries give money to poor countries? They also included issues of intergenerational justice: What are the obligations of the present to the future?

In academic circles, these issues were intensely debated, and before I joined the Obama administration, I had been working on a book on them.

The White House asked me to abandon the book, which I was coauthoring with Eric Posner and David Weisbach at the University of Chicago, on the ground that it said a few things that did not align with the policies of the Obama administration. That was pretty maddening, but I understood. A public official, working for the government, ought not to embarrass his boss. Posner and Weisbach published the book without me.[3] I admire their book, and I was in fundamental agreement with it back in 2009. I still admire the book, but with fear and trembling (Posner and Weisbach are never wrong), I am now in fundamental disagreement with it, for reasons explained in chapter 2. To get ahead of the story, I now think that arguments about corrective justice and distributive justice, calling for a significant transfer of resources from wealthy, high-emitting countries to poor, low-emitting countries, are essentially right. Those arguments do run into an assortment of formidable objections, but still, they are essentially right.

Returning to 2009: It turned out that the academic debates bore directly on intensely practical issues. Greenstone and I ended up cochairing a technical working group on the social cost of carbon, and it took many months for the group's members to get consensus (more on all this in chapter 1). It was, I think, the hardest project I have ever been involved with (substantively difficult, exhausting), and I have been haunted, ever since, by the underlying ethical questions. I don't think a week has gone by without my thinking about them.

I left the Obama administration in 2012, but I have continued to be involved in climate change issues, both inside and outside the United States government, most recently as cochair of the Climate Change Action Group in the Department of Homeland Security. A key lesson is that ethical issues are closely intertwined with *strategic* issues. When rich countries ask poor countries to do more to reduce their emissions, poor nations might well say: "If you want us to do that, you will have to pay us." Or: "You caused the problem, and got rich in the process, and now you want us to solve the problem? Wow. How much is that worth to you?"

Rich countries might well respond: "We are all in this together, and you are more at risk than you think. Actually, you are more at risk than we are. Let's find a reasonable path, without your demanding massive subsidies, which will derail the whole effort!"

We are going to bracket question of strategy and negotiation here. (Well, almost entirely.) We will focus on ethical questions: Who owes what to whom?

## FINLEY

To get a grip on those questions, imagine that the world consists of just one person, named Finley. As it happens, Finley is an innovator, and he has invented a lot of things. Some of his inventions are making his life much better, but they also emit certain pollutants, called greenhouse gases, which are making the world warmer and the weather more volatile. The changes are gradual. It appears that things will get worse for Finley in about a decade, and much worse for him in fifty years, and much, much worse for him in a hundred years.

The good news is that unless something terrible happens to him, Finley is likely to live for an immensely long time. He might be immortal. Even if he is not, he has reason to believe that he might well live to three hundred, or four hundred, or perhaps even more. (If you are thinking that Finley is meant as a stand-in for humanity, well, I won't deny it.)

What should Finley do? It goes without saying that he ought to ask whether he should act to reduce his greenhouse gas emissions. It is relevant whether his life will get worse if he does that, and how much worse it will get.

Let us suppose that if Finley immediately cuts his emissions in half, his life will get worse in meaningful ways. He will have to scale back. He will have to consume less. Before very long, he might be able to innovate and to restore the status quo, but he is not at all sure that he can do that. At the same time, he is acutely aware that if he continues with "business as usual," his emissions will make his world a lot hotter, and possibly a lot more dangerous, in a few decades.

Finley might well embrace a simple framework to think about how to proceed: He should act to maximize his well-being over his (very long) lifetime. He might well think that his future self matters every bit as much as his current self; he can find no reason to think that his self, right now, deserves more concern than his self in a hundred years. In short: Finley Now does not deserve more attention and concern than Finley Later. He knows, of course, that there is some chance that he will die, and he takes that into account, but he is feeling great, and he is pretty confident that he will live for a very long time.

Finley applies the golden rule to his various selves, extending over time. He follows a principle of neutrality with respect to each and every Finley, at each and every moment. It follows that a terrible decade many decades hence is not less bad than a terrible decade after this one.

Finley is not sure whether to be a utilitarian because he is not sure what *utility* is or what it includes. But because he cares about his own welfare, he is willing to be a *welfarist,* who seeks to maximize his welfare over the course of his lifetime. Finley thinks that the idea of *welfare* is more capacious than the idea of *utility*, which is often thought to focus on *happiness*—or more narrowly, pleasure and pain. For Finley, welfare includes a sense of meaning or purpose, not merely happiness.[4] But Finley does not want to fuss much over philosophical issues. He simply wants to insist that he cares a lot about his welfare, not only today but also tomorrow, and the day after that as well.

Finley's welfarism has concrete consequences for his conduct. He knows that he cannot eliminate his greenhouse gas emissions today, tomorrow, or the next day. But he also knows that he has to find a way to scale back. How much should he scale back? The answer depends on what is optimal for him, given his commitment to welfarism. Finley knows, too, that he might be able to adapt to what appears to be coming. He might be able to take steps to reduce the risks. He has to make choices between *mitigation*, coming from reductions of emissions, and *adaptation*, coming from steps to ensure that a warmer world is not a more dangerous world.

## FINLEY AND MAYA

Now suppose that the world consists of two people: Finley and Maya.

While Finley emits a lot of greenhouse gases, Maya does not. While Finley has plenty of material goods and loves his life, things are much worse for Maya. She has little in the way of material goods; some days are horrible for her, and though some days are good, she is keenly aware that her life could be a lot better than it now is. Because Finley emits such a high volume of greenhouse gases and occupies the same world as she does, Maya is at grave risk. Maya thinks that Finley is endangering her.

Like Finley, Maya has reason to think that she will live an immensely long time. She might be immortal; she has reason to believe that at the very least, she will live for hundreds of years. To this extent, she is like Finley. In these circumstances, she wants to do two things. First, she wants to live in a world that does not become unbearably hot. Second, she wants to have more in the way of material goods, and to do that, she has to engage in a host of activities that involve emissions of greenhouse gases.

What should Finley do? What should Maya do? It does not appear that Maya has a great deal of bargaining power. But she can tell Finley that unless he cuts back his emissions, and finds a way to protect her from unbearable heat, she will not cut back her own. That will in turn endanger Finley.

Finley might be selfish. He might think: *I care about my own welfare, not Maya's, and I will proceed in such a way as to promote my own welfare.* If so, we could imagine a difficult bargaining situation. What will (or can) Finley offer Maya? What can (or will) Maya offer Finley?

But following Mill, Finley might conclude that selfishness is unethical. Maya counts too. Why should Maya count less than Finley does? On reflection, and as a matter of principle, Finley should be tempted to think: *I am a welfarist. I want to maximize the welfare of Maya and myself. The two of us count equally. I am willing to scale back my emissions to the point where further reductions would cause the two of us more harm than good. If that means that I scale back only a little, fine, and so too if that means that I scale back a lot.*

Finley might also be tempted to think: *I would be willing to support an agreement with Maya by which we both agree to scale back our emissions to the point where further reductions would cause the two of us (each counted equally) more harm than good.*

Maya might agree that all this is a good start, but she might not be at all satisfied. She might insist that Finley is responsible for the problem in the first place. After all, he has become prosperous at her expense. She thinks that he has wronged her—and that he should both pay her for the damage and stop doing it.

She might also insist that Finley is doing very well in terms of both prosperity and welfare, and that she is doing much less well. She might argue that because he is rich and she is not, he is the one who should bear all or almost all of the cost of emissions reductions, and of adaptation to what is here and getting worse. She might urge that because she is struggling in life, and much worse off than Finley is, she deserves priority. She is a *prioritarian.*

## BACK TO REALITY

With respect to the real world of climate change, you might think of Finley as analogous to the United States and other wealthy nations (and perhaps China too) and of Maya as analogous to India, Afghanistan, the Central African Republic, Sudan, and other poor or poorer nations. I offer the tale of Finley and Maya to provide a glimpse into three ways to think of some of the ethical issues raised by climate change. One perspective is *welfarist*: we ought to maximize human welfare, treating everyone essentially the same, and welfarism should be the foundation for analysis.

A second perspective points to *fairness*: those who have harmed others should pay for the harm they have caused. This is an idea about corrective or compensatory justice. As we shall see, there might well be welfarist reasons to focus on fairness, because focusing on fairness might turn out to increase welfare.

A third perspective is *distributional*: people who are wealthy should be willing to give to those who are not. As we shall also see, there are welfarist reasons to focus on fair distribution. If, for example, the wealthy Finley loses some material goods and the poor Maya gets them, Finley might be only a little worse off (in terms of welfare), and Maya might be a lot better off (in the same terms). Independent of that point, we might be prioritarians, insisting that those at the bottom deserve priority. (Prioritarianism will come up at various places in this book, so please remember the term and the basic concept.)

Nations are not people, of course. An analogy is not an identity. The United States is not Finley. It has well over 330 million people, many of whom are not rich, and many of whom have contributed little or nothing to the problem of climate change. India is not Maya. It has over one billion people, many of whom are rich, and many of whom are responsible for significant emissions. These points matter.

For the most part, welfarism will provide the foundations for the analysis in this book, and the golden rule will play a prominent role. It follows that the form of welfarism that I will be endorsing counts each person equally, regardless of where they live and when they live. We shall see that if the world seeks to solve the climate problem, each country needs to consider the interests of people who live in other countries—and that if each country does not do that, we are all in big trouble.

We shall see that distributive justice and corrective justice greatly matter, but that there are real complications in applying those ideas to the international context. We shall also see that a person alive today is entitled to no more consideration than a person born decades from now—which means that we should endorse a principle of *intergenerational neutrality*. (Still, we should be discounting money, which turns out to be important.)

We shall explore the monetary valuation of climate-related risks. Unpleasant though it might seem, monetization of such risks is essential; we cannot decide what to do about climate change without monetization, even it is only implicit (and thus not transparent). We shall see that it is imperative to focus on the challenge of adaptation, and that we already can specify the foundations of policies that would promote or enable adaptation. We shall see, finally, that consumers and consumer choices are close to the heart of the problem of climate change, and that Choice Engines of various sorts might at once improve consumer welfare and reduce emissions. To show this, we will say something about algorithms, artificial intelligence (AI), and large language models (LLMs).

A word on exposition before we begin: Some of the issues here are highly technical, in areas that involve not only the science of climate change, but also economics, psychology, and law. In some places, it will be necessary to write with a degree of formality and to use some technical language. In such places, I fear, the discussion might be a bit of a slog. But the underlying concepts are, I hope, intuitive and reasonably straightforward, and I have tried to avoid excessive formality here.

# 1 CLIMATE CHANGE COSMOPOLITANISM

My central question in this chapter is whether a nation, such as the United States, France, Denmark, Egypt, or Germany, should take account of the harms it does to people in other nations when deciding whether to scale back its own greenhouse gas emissions. Some people think that a nation should focus only on the harms that it does to its own citizens, or those within its own borders. I will be rejecting that view and arguing in favor of a form of *climate change cosmopolitanism*. I will be arguing that climate change cosmopolitanism is justified on moral grounds. I will also be arguing that climate change cosmopolitanism is justified by reference to domestic self-interest.

But before we get there, we have to clear some ground.

## BEYOND SCIENCE

Is climate change a problem of science? Is it a problem of politics? Is it a problem of ethics? What kind of problem is it?

Many people insist that it is essentially a problem of science. They say that we should "follow the science." Other people answer that science is inconclusive on important questions. There is so much that we do not know; the whole domain, they insist, is full of guesswork and speculation. They add that even if science gives clear answers about the effects of greenhouse gas emissions, the question of what to *do* is inevitably a political one, which means that it is also an ethical one. For example, an understanding of the science cannot possibly tell us whether the United

States and Europe should compensate poor nations for the harms done by climate change. Nor can an understanding of science tell us what people today owe to future generations. If science tells us that people one hundred years from now will be highly vulnerable, the question remains: What are our ethical obligations?

The debate over the role of science, politics, and ethics in climate policy is connected with a much broader struggle between technocratic and political conceptions of governance, law, and regulation. On one view, regulatory decisions, including those involving climate change, call for a high degree of technical expertise. We need technocrats, not politicians. We might even need algorithms and artificial intelligence, not human beings. An important implication is that much of the time, the political convictions of current political leaders should not much matter. Facts are what matter. Accurate predictions are what is needed. It follows that across different administrations, with different political convictions, continuity must be maintained, at least if technical expertise calls for a particular approach. Something like this might be true in the United States, the United Kingdom, Ireland, France, Germany, Mexico, and the world over.

The technocratic view might sound odd and jarring, but my own experience inside the U.S. government suggests that it has more than a grain of truth—much more, in fact, than one might think. If the Environmental Protection Agency is aiming to protect people from ozone and particulate matter, for example, science offers a lot of guidance. Some decisions on those subjects would be fatally inconsistent with science because they would be too weak to protect public health. Other decisions would be inconsistent with science because they would be far too aggressive and would go well beyond what science suggests is necessary to protect public health.

The same can be said about climate change. It would be preposterous to say that greenhouse gas emissions do not endanger public health and welfare. Some claims about the extraordinary risks posed by current emissions are also preposterous; they go far beyond what science says.

But there is a different view, captured in a simple phrase: *elections have consequences*. In that view, decisions with respect to climate change and other problems do and should depend on political convictions, broadly understood. When I worked in the Obama and Biden administrations, we issued a number of regulations involving climate change, designed to reduce emissions from appliances, motor vehicles, and power plants. The Trump administration undid or reversed many of the regulations issued in the Obama administration. The Biden administration restored many of them, or acted more aggressively than the Obama administration did. Something similar can be said about nations all over the world. In Europe, for example, shifts in directions have been common and occasionally dizzying.

With examples like this in mind, many people think that across administrations, continuity need not be maintained. Facts are hardly all that matter. With respect to health, safety, and the environment, the Biden administration need not accept the judgments of the Trump administration, and the Trump administration need not accept the judgments of the Obama administration. A page of history is worth a thousand pages of analysis.

In my view, these claims are not altogether wrong, but they are too simple. Both technocratic and political conceptions of governance, law, and regulatory practice are crude. Some regulatory judgments depend on broadly political convictions, and that is entirely appropriate. In Europe, Africa, and North America, for example, different administrations need not approach climate change, air pollution, or occupational safety in the same way.

At the same time, elections do not and should not *always* have consequences. Government agencies cannot say that two plus two equals five, that benzene is not a carcinogen, that smoking does not cause lung cancer, or that greenhouse gas emissions do not cause climate change. The question whether and to what extent regulatory decisions may be, and should be, technocratic or political depends on the context, including the text of the laws enacted by the national legislature and existing evidence. The question requires answers to a host of subsidiary questions.

As William Blake wrote, commenting on Sir Joshua Reynolds: "To Generalize is to be an Idiot[.] To Particularize is the Alone Distinctive of Merit."[1] True, Blake generalized there, but let us not be fussy.

## THE MOST IMPORTANT NUMBER YOU'VE NEVER HEARD OF

What is the appropriate stringency of laws and regulations designed to reduce greenhouse gas emissions? How should governments decide whether to reduce emissions modestly, aggressively, or much more than that?

The initial answers, and potentially the decisive ones, depend, of course, on the relevant law. What does it require or permit? In the United States, moreover, Executive Order 12866[2] (issued by President Bill Clinton) and Executive Order 13563[3] (issued by President Barack Obama) govern regulatory decisions, including those that involve climate change. Under those orders, government agencies in the United States are required to quantify the costs and benefits of their regulations, and to proceed, to the extent permitted by law, only if the benefits justify the costs.

That sounds a bit technical, and it is, but we are going to avoid most of the technicalities here. You can raise a lot of questions about cost-benefit analysis, in general and in the context of climate change.[4] Let us just note that many people believe that an analysis of costs and benefits is essential if we are trying to decide on whether and how stringently we should regulate. (I share that belief.) The reason is that unless we know something about costs and benefits, we will not know nearly enough about whether and when regulation will improve social welfare or make human lives better. (As I have said, the lives of nonhuman animals matter as well, and what happens to those lives should be part of the assessment of what would improve social welfare.)

A climate change regulation that costs a lot and delivers modest benefits is not a good idea. A climate change regulation that costs very little and that benefits a lot of people (and nonhuman animals) is an excellent idea. With respect to climate change, we ought to focus on the consequences of what we do—on whether we are addressing the problem or not. We ought to focus much less on whether what we do "makes a statement." The real question is what the statement does in the world, not what kind of statement it is.

## NUMBERS

What are the benefits and costs of efforts to reduce greenhouse gas emissions?

It would be preposterous to say that as of tomorrow, the level of such emissions must drop to zero. Given the magnitude of the climate change problem, it would also be preposterous to continue with "business as usual" and do nothing at all to reduce such emissions. Many people involved in climate change discussions fervently urge the world to adopt a ceiling—perhaps 2°C, perhaps 1.5°C. I will have more to say about such ceilings (and I will not be so enthusiastic about the idea), but notice for now that they claim to be rooted in science and also that they do not by themselves specify the benefits of particular measures that reduce emissions by identifiable amounts.

Sometimes described as "the most important number you have never heard of,"[5] the social cost of carbon plays a major role in climate change policy.[6] In brief, the social cost of carbon is meant to identify the monetary cost of a ton of carbon emissions and thus to help specify the benefits of efforts to reduce such emissions. It is a matter of simple arithmetic: if the social cost of carbon is $100, then a reduction of five hundred tons of carbon would produce monetized benefits of $50,000.

It is important to note that in the United States, regulators now speak of the social cost of greenhouse gases, not just the social cost of carbon.[7] Methane and nitrous oxides are included among the greenhouse gases whose social cost is being monetized. For the sake of simplicity, I shall refer to the social cost of carbon; broadly speaking, the analysis is essentially the same for other greenhouse gases.

The social cost of carbon has had a massive impact on many nations, including the United States. In that country, it was used in eighty-three regulatory or planning processes, from six agencies, from 2009 to 2016.[8] It has been used on numerous occasions from 2017 to the present. A high social cost of carbon will of course tend to support aggressive climate change regulations, while a low one will tend to support modest regulations. But how should public officials, and reviewing courts, decide whether the number should be high or low? How should they choose a number?

It is tempting to think that the ultimate judgment is political, but we have seen that the temptation should be rejected.[9] The law matters, and so does the science. With respect to scope and levels of stringency, legislative enactments constrain agency judgments, and within the boundaries set by those enactments, agency decisions must not be arbitrary. With respect to arbitrariness in particular, and sound policy more generally, the social cost of carbon depends on a host of scientific or technical conclusions, or at least on conclusions with heavy scientific or technical elements.

Consider just four issues:

- It is necessary to assess *climate sensitivity*.[10] For a given level of worldwide emissions, what is the likely change in temperature? How sensitive is the climate to increases in emissions?

- It is necessary to assess the *damage function*. For a given change in temperature, what are the likely effects on human welfare?[11] If the world becomes 1.8°C hotter, how will people be affected? What would happen instead if the world became hotter by 1.9°C, or 2.5°C, or 3°C?
- It is necessary to produce an appropriate discount rate (taken up in chapter 3). How should we treat future harms and future benefits? If a wide range of climate-related harms (monetary losses, deaths, injuries) would occur in 2100, should they be discounted at a rate of 1 percent, 2 percent, 3 percent, or more?
- It is necessary to consider the issue of adaptation (taken up in chapter 5). If global temperatures rise by (say) 1.8°C by 2100, how will human beings adapt? If they will do a great deal to adapt and are able to do so at relatively low cost, then the social cost of carbon is lower than if they will do very little to adapt. If serious adaptation will not occur, or if it will not avert horrific harm, then the social cost should not be much lowered.

None of these questions has an obvious answer. And in terms of policy, there is another question, my focus here. It is a fundamental one that can, in principle, be separated from the rest: Should a nation use a global measure of damages, or should it use a domestic measure?[12] Let us focus on the United States in particular, noting that the analysis would not be very different if we were speaking of Canada, France, Denmark, Germany, South Korea, South Africa, or China.

Use of the global number would reflect what we might describe as a form of *cosmopolitanism*, justified on moral, strategic, or other grounds. The reason is that by definition, a global measure would include the damage done by U.S. greenhouse gas emissions to people all over the world.[13] By definition, a domestic measure would include damages only within (or to) the United States.

As we shall see, the best understanding of the domestic measure raises some exceedingly hard questions of fact. But on any account, the domestic measure is some fraction of the global measure. For example, the difference between the measure under the Trump administration (about $2) and that under the Obama and Biden administrations (about $50 under Obama and also about $50 until late 2023 under Biden, with a figure of about $190 beginning in December 2023) turns in large part on the fact that the former is a domestic measure, while the latter two are global measures. (The Trump administration also chose a higher discount rate, which made present damages significantly lower—but the main driver of the difference between the Trump administration on the one hand and the Obama and Biden administrations on the other is the choice of the domestic number for the former and the global number for the latter.)

My central goal in this chapter is to explore the choice between the two, assuming that the relevant law allows government agencies to decide as they (reasonably) see fit. As we shall see, there are four main arguments in favor of using the global figure. In brief, they take the following form:

1. The *epistemic* argument: We do not know a great deal about the purely domestic harms likely to come from climate change, let alone from different incremental increases in greenhouse gas emissions. At the very least, we cannot easily specify the purely domestic harms. Our lack of knowledge makes it difficult or perhaps even impossible to generate a purely domestic number.
2. The *interconnectedness* argument: The world is interconnected, and many Americans live abroad. Harms done by domestic emissions to Americans and American interests are hardly limited to those done by weather-related events within in the United States. Among other things, they include (1) harms to U.S. citizens living abroad and (2) economic and other harms to U.S. citizens that come from the cascading effects of harm done to foreigners (including governments, companies, and individuals). Any approach to the domestic harms done by climate change must take account of points 1 and 2 (and other relevant harms). This point of course holds for all nations, not just the United States.
3. The *moral cosmopolitan* argument: In deciding on the scope of its regulations, the United States should take into account the harms it does to non-Americans. This is so for emphatically moral reasons, signaled by the golden rule and utilitarianism, and their moral foundations.[14] This point also holds for all nations.
4. The *reciprocity* argument: If all nations used a domestic figure, all nations, including the United States, would lose. Use of the global figure, by all nations, is very much in the interest of the United States. A successful approach to the climate change problem requires nations to treat greenhouse gas emissions as producing global, and not merely domestic, externalities. According to the reciprocity argument, international leadership

by the United States is an important and perhaps even essential means of creating a norm, and imposing incentives, to overcome a (wicked) prisoner's dilemma. The reciprocity argument can be seen as a form of strategic or backdoor cosmopolitanism; it insists on taking account of harms done to foreigners, not because doing so is morally required, but because doing so is in the nation's purely domestic self-interest.

My basic conclusions are straightforward:

1. The premise of the epistemic argument is entirely correct, but by itself, the argument cannot possibly justify the use of a global number. In any accounting, the global number is too high if the concern is the harm done to U.S. citizens and U.S. interests. It remains true that specification of the domestic number is exceptionally difficult, but it simply cannot be equivalent to the global number. For the domestic number, evidence might be consulted, and some rules of thumb might be justified, to develop lower and upper bounds in the face of the epistemic challenges.
2. The interconnectedness argument is also correct, and it justifies use of what might be called an *inclusive domestic number*. That is important. But by itself, the incompleteness argument does not justify choice of the global number. In any account, the domestic number, even if inclusive, must fall short of that number.
3. The moral cosmopolitan argument is very strong. Unless the law forbids its use (an important question), it is generally convincing, at least in the context of harm done by people in one nation to those in another nation. This is so whatever we think of cosmopolitanism in the abstract or more generally. The challenge is that moral cosmopolitanism might seem to have large and perhaps unacceptable implications for national policy in general: Is the United States really under an obligation to value the lives of noncitizens as much as it values the lives of its own citizens? With respect to military conflict or foreign aid? The moral cosmopolitanism argument would be most convincing if it can be narrowed and specified.
4. The reciprocity argument is also very strong, though it runs into some serious complications and some counterarguments that must be engaged. The most fundamental objection is that because we are dealing with a prisoner's dilemma, nations will have a strong motivation to defect. To say the least, *a decision by the United States to use the global*

*figure will not automatically lead the rest of the world to do the same.* Why should the United States impose costs on itself when other nations are not bound to follow, and have every incentive to stand on the sidelines and take advantage of its actions? The best answer points to international leadership on the part of the United States (a form of "leading by example"), and to reasonable conjectures about how international bargaining might be expected to unfold. Even if those conjectures are merely that, using the global number is appropriate in light of the sheer magnitude of the stakes. I am aware that this is a compressed argument; I shall offer more details in due course.

## COUNTING FOREIGN LIVES?

The question whether, when, and how foreign lives (and interests) should be counted in national policymaking raises fundamental issues in many fields, including political philosophy, political science, ethics, economics, and law. On one view, apparently compatible with the golden rule and utilitarianism, foreign lives should be counted for their own sake. A life is a life, after all, and perhaps a foreign life should be given no less attention, in principle, than a domestic life. Why should the life of a person born in the Central African Republic be worth less than the life of a person born in the United States?[15] From the moral point of view, it is far from clear that the value of a life is less, or falls to zero, because of where one lives.

On a more cautious approach, foreign lives should be counted because and to the extent that Americans believe that they should. Perhaps government agencies should ask that question, and perhaps they should be interested in whether and to what extent Americans are willing to pay to protect those lives or to reduce statistical risks to them.[16] It is highly likely that Americans are willing, on average, to pay *something*, even if they are not willing to pay as much as they are willing to pay to protect Americans. (If you think the idea of willingness to pay is odd or misplaced in this context, you might be right. But it is a conventional approach, and I will have more to say about it in chapter 4.)

On one view, then, foreign lives matter simply because they do, on principle. On another view, foreign lives matter to the extent that Americans believe that they do (and are willing to back that belief with money). There is a third view: Foreign lives matter because and to the extent that a decision to consider them is in the strategic interests of the United States. If Americans care about foreign lives, foreign lives (or foreign nations) will care about American lives, and that is good for Americans. Suppose, for example, that foreign aid is undertaken with the goal of saving foreign lives. That may or may not be simple altruism. The United States and its citizens might be beneficiaries, certainly in the long term and perhaps in the short term as well.

Nations need friends. Friendship is a good thing, especially in a world in which nations compete with one another for alliances. Climate change is an exceedingly good domain in which to forge and cultivate friendships. How much should a nation do to cultivate friendships? That question does not have an obvious answer. Let us simply say: a lot.

It should be clear that the United States must ask whether and how much foreign lives matter in many contexts, regulatory and nonregulatory. The military context is obviously relevant: If the United States can prevent ten thousand foreign deaths at the cost of three hundred American lives, should it do so? What if the corresponding numbers are fifty thousand and twenty-five, or four thousand and one thousand? What if the prevention of those foreign deaths reduces security risks to the United States? What if the prevention of those foreign deaths creates beneficial or even essential alliances?

Similar questions might be asked in the context of foreign aid. Suppose that the United States can save one hundred foreign lives at a domestic cost of $900 million. Should it do so? What if the corresponding numbers are ten thousand and $200 billion, or ten and $700 million? Once again: What if the prevention of those foreign deaths creates friendships? What if it reduces security risks to the United States? What if the prevention of those foreign deaths creates beneficial alliances? Starting in 2022, all of these questions became highly relevant, of course, in the context of the war waged by the Russian Federation on Ukraine.

The regulatory context might be thought to be very different from that of foreign aid. If the United States prevents harm to another nation, it might not be giving a gift. At least this is so if the United States is seeking to ensure that (say) its own companies are not harming those outside of its territorial borders. If so, the United States is seeking to avoid commission of something like a wrong or a tort, or maybe even a murder; it is not seeking to confer a benefit or to practice a kind of charity.

## THEORY AND PRACTICE

To understand government practice, we now need to get a bit technical. To decide whether to use a global figure or a domestic figure for the social cost of carbon, the initial question is one of law, and it will differ from statute to statute and from domain to domain. The question of air pollution regulation may be different from the question of water pollution, and both are likely to be different from that of foreign aid, and all three may differ from the question of national security policy. With respect to regulation, we can imagine three possibilities under relevant statutes:

1. An agency is required to consider global impacts.
2. An agency is required to consider only domestic impacts.
3. An agency is permitted but not required to consider either global or only domestic impacts, which means that it gets to exercise discretion.

There is no reason to think that the answer under one statute will be the same as the answer under another statute. For regulation in general, guidance from the Office of Management and Budget (OMB) for agencies conducting cost-benefit analyses has long leaned toward domestic effects: "Your analysis should focus on benefits and costs that accrue to citizens and residents of the United States. Where you choose to evaluate a regulation that is likely to have effects beyond the borders of the United States, those effects should be reported separately."[17]

The statement "should focus" seems to suggest that domestic effects are what matter. Interestingly, however, those domestic effects include benefits and costs with respect to residents and not merely citizens, which suggests that to that extent, foreign lives *do* matter (at least if they are lived within the United States). Also interestingly, agencies are not merely authorized but also directed ("should") to report extraterritorial effects ("separately"), though OMB does not indicate what weight, if any, ought to be given to them.

No comprehensive account of actual regulatory practice exists, but Arden Rowell and Lesley Wexler have proved some valuable glimpses.[18] They find that at a single point in time (August 2011), the seven largest regulations could reasonably be expected to reduce mortality risks in foreign nations—but that *in none of them did agencies attach any value at all to foreign lives.*[19]

For example, certain air pollution regulations from the Environmental Protection Agency (EPA) might be expected to have beneficial effects in Canada and Mexico, but the EPA did not attempt to catalog those effects. The relevant regulatory impact analyses appear entirely to have excluded populations outside of the United States, which suggests that the EPA did not value mortality benefits to foreigners at all. Arden and Wexler conclude that "the impact of agencies' implicit practices is to value foreign lives at zero; that is, to commit *no* domestic resources towards the protection of foreign lives."[20]

To say the least, that approach is not consistent with cosmopolitanism, or with the golden rule.

## A LITTLE HISTORY

The development of a social cost of carbon by the U.S. government has a lengthy history, with some major changes in direction.[21] For present purposes, the central point is that all four arguments in favor of the use of a global SCC have been explicitly invoked by the executive branch, which is to say that agencies have been keenly aware of the theoretical arguments and their relevance to practice.

To make a long story short: both the Obama and Biden administrations used all four arguments, though with different emphases at different times. The Trump administration implicitly rejected all of them, but without much of an explanation, which created serious trouble in court.[22]

### THE OBAMA ADMINISTRATION

The initial guidance, coming in the form of a technical support document (TSD), was issued by the Interagency Working Group (IWG) on the Social Cost of Carbon in 2010.[23] The IWG was headed by the Council of Economic Advisers (CEA), originally in the person of Michael Greenstone, and the OMB, originally in the person of the present author.

The IWG included a wide array of agencies: CEA, the Council on Environmental Quality, the Department of Agriculture, the Department of Commerce, the Department of Energy (DOE), the Department of Transportation, the Environmental Protection Agency, the National Economic Council, the Office of Energy and Climate Change, OMB, the Office of

Science and Technology Policy, and the Department of the Treasury. All of them contributed; the process was lengthy, intense, and substantive (and not at all political). The resulting document describes the monetary value of reductions in carbon emissions. In 2010, the United States did, in a sense, put a price on carbon.

Crucially, the TSD chose the global rather than merely domestic measure of damages; harms from U.S. emissions to people in China, Europe, Africa, India, and elsewhere are counted.[24] At various stages during the Obama administration, that issue received a great deal of attention, with reference to all four of the arguments I have sketched.

In its initial statement in 2010, the TSD noted that climate change involves "a global externality" and that it "presents a problem that the United States alone cannot solve."[25] By itself, of course, the fact that a global externality is involved is not sufficient to justify use of a global number. But the IWG added that "the United States has been actively involved in seeking international agreements to reduce emissions and in encouraging other nations, including emerging major economies, to take significant steps to reduce emissions."[26] In that way, it seemed to signal the reciprocity argument, and perhaps also to gesture in the direction of moral cosmopolitanism.

In 2016, the IWG offered a more detailed account. It acknowledged that under OMB Circular A-4 as it existed at the time, "analysis of economically significant proposed and final regulations from the domestic perspective is required, while analysis from the international perspective is optional."[27] At the same time, it concluded "that a modified approach is more appropriate in this case."[28]

To defend that conclusion, it made several points, in a way that seemed to meld together the four arguments I am exploring here. Climate change, the IWG reiterated, "involves a global externality: emissions of most greenhouse gases contribute to damages around the world even when they are emitted in the United States—and conversely, greenhouse gases emitted elsewhere contribute to damages in the United States."[29] It follows that climate change is a problem "that the United States alone cannot solve. Other countries will also need to take action to reduce emissions if significant changes in the global climate are to be avoided. . . . Using a global estimate of damages in U.S. regulatory analyses sends a strong signal to other nations that they too should base their emissions reductions strategies on a global perspective, thus supporting a cooperative and mutually beneficial approach to achieving needed reduction."[30] This is the reciprocity argument in action.

In addition, the adverse impacts of climate change "on other countries can have spillover effects on the United States, particularly in the areas of national security, international trade, public health, and humanitarian concerns."[31] The IWG also made a technical point about the limits of existing research, and thus made the epistemic argument. As "an empirical matter, the development of a domestic SC-GHG is greatly complicated by the relatively few region- or country-specific estimates of the SC-$CO_2$ in the literature"; existing estimates are incomplete.[32]

Finally, and importantly, the problem of climate change presents a prisoner's dilemma: if every nation used a domestic cost of carbon, every nation would be hurt.[33] This is plainly "the issue of reciprocity."[34] It follows that the "only way to achieve an efficient allocation of resources for emissions reduction on a global basis is for all countries to base their policies on global estimates of damages."[35]

## THE TRUMP ADMINISTRATION

In 2017, President Donald Trump issued an executive order that explicitly addressed the social cost of carbon.[36] The order was not exactly cautious. It briskly disbanded the IWG and rescinded essentially all relevant documents from the Obama administration. In addition, it stated:

> Effective immediately, when monetizing the value of changes in greenhouse gas emissions resulting from regulations, including with respect to the consideration of domestic versus international impacts and the consideration of appropriate discount rates, agencies shall ensure, to the extent permitted by law, that any such estimates are consistent with the guidance contained in OMB Circular A-4 of September 17, 2003 (Regulatory Analysis), which was issued after peer review and public comment and has been widely accepted for more than a decade as embodying the best practices for conducting regulatory cost-benefit analysis.[37]

This provision was widely understood to call for two fundamental changes to the Obama administration's practices. First, agencies would be expected to use the domestic measure rather than the global measure. Second, agencies would be expected to calculate the benefits of reducing greenhouse gas emissions with discount rates of 3 percent and 7 percent, consistent with

OMB Circular A-4 as it existed at the time (see chapter 3). The upshot is that in regulations from the Trump administration, the social cost of carbon generally ranged from one dollar to seven dollars—a small fraction of the number that would come from the analysis of Obama's IWG (approximately fifty dollars).

## THE BIDEN ADMINISTRATION

On his very first day in office, President Joe Biden issued an executive order that, among other things, explicitly addressed the social cost of carbon.[38] The relevant provision began simply: "It is essential that agencies capture the full costs of greenhouse gas emissions as accurately as possible, including by taking global damages into account." This was a clear direction to use the global rather than the domestic measure.

The order added that taking those steps "facilitates sound decision-making, recognizes the breadth of climate impacts, and supports the international leadership of the United States on climate issues." It also established a new Interagency Working Group, led by the Council of Economic Advisers, the director of OMB, and the director of the Office of Science and Technology Policy. The working group was directed to produce an interim SCC, social cost of nitrous oxide (SCN), and social cost of methane (SCM) within thirty days. It was also directed to "publish a final SCC, SCN, and SCM by no later than January 2022."

The interim social cost of carbon (actually, greenhouse gases) adopted much of the analysis and approach of the Obama administration, at least for that interim use.[39] At the same time, it offered significant discussion of relevant issues, including the choice between the global and domestic measures. With respect to the use of the global measure, it said:

> Unlike many environmental problems where the causes and impacts are distributed more locally, climate change is a true global challenge making GHG emissions a global externality. GHG emissions contribute to damages around the world regardless of where they are emitted. The global nature of GHGs means that U.S. interests, and therefore the benefits to the U.S. population of GHG mitigation, cannot be defined solely by the climate impacts that occur within U.S. borders. Impacts that occur outside U.S. borders as a result of U.S. actions can directly and indirectly affect the welfare of U.S. citizens and residents through a multitude of pathways. Over 9 million U.S. citizens lived abroad as of 2016–17 and U.S. direct investment positions abroad totaled nearly $6 trillion in 2019.

> Climate impacts occurring outside of U.S. borders will have a direct impact on these U.S. citizens and the investment returns on those assets owned by U.S. citizens and residents. The U.S. economy is also inextricably linked to the rest of the world. . . . The global nature of GHGs means that damages caused by a ton of emissions in the U.S. are felt globally and that a ton emitted in any other country harms those in the U.S. Therefore, assessing the benefits of U.S. GHG mitigation activities will require consideration of how those actions may affect mitigation activities by other countries since those international actions will provide a benefit to U.S. citizens and residents.[40]

This analysis emphasizes (and specifies) the interconnectedness argument and also nods in the direction of the reciprocity argument.

In 2022, the EPA issued a detailed draft report in which it built on these arguments.[41] The EPA reiterated that "GHG emissions are a global externality making climate change a true global challenge." Elaborating the interconnectedness argument, the EPA added that "climate change will directly impact U.S. interests that are located abroad (such as U.S. citizens, investments, military bases and other assets, and resources in the global commons [e.g., through changes in fisheries' productivity and location])." Almost 40 percent of the equity holdings of U.S. pension plans are in foreign stocks, and the United States has more than five hundred military installations in forty-five foreign nations. "Climate impacts that occur outside U.S. borders will impact the welfare of individuals and the profits of firms that reside in the U.S. because of their connection to the global economy."

There are also issues of international migration. The EPA noted that "robust estimate of climate damages to U.S. citizens and residents that accounts for the myriad of ways that global climate change reduces the net welfare of U.S. populations does not currently exist in the literature." This is the epistemic argument. The EPA also emphasized the reciprocity argument: what the United States does is likely to affect what other nations do. That is important because "the only way to achieve an efficient allocation of resources for emissions reduction on a global basis—and so benefit the United States and its citizens and residents—is for all countries to consider estimates of global marginal damages."

In 2023, the EPA finalized its report, building on the draft and again offering an elaborate treatment of the use of the global number.[42] Once more it stated that "GHG emissions are a global externality making climate change a true global challenge. GHG emissions contribute to damages around the world regardless of where they are emitted." With respect to interconnectedness, it noted that "U.S. interests are affected by climate change impacts through a multitude of pathways, and these need to be considered when evaluating the benefits of GHG mitigation to the U.S. population." Among other things, it added that "the U.S. has over 500 military sites abroad across 45 foreign countries."[43] It elaborated:

> The U.S. economy is also inextricably linked to the rest of the world. The U.S. exports over $2 trillion worth of goods and services a year and imports around $3 trillion. According to recent data, over 20% of American firms' profits are earned on activities outside the country. Climate impacts that occur outside U.S. borders will impact the welfare of individuals and the profits of firms that reside in the U.S. because of their connection to the global economy. This will occur through the effect of climate change on international markets, trade, tourism, and other activities. Supply chain disruptions are a prominent pathway through which U.S. business and consumers are, and will continue to be, affected by climate change impacts abroad.[44]

The EPA also pointed to the epistemic challenge: "The global models used in SC-GHG estimation do not lend themselves to disaggregation in a way that can provide a comprehensive estimate of climate change damages to U.S. citizens and residents that accounts for the myriad of ways that global climate change reduces the net welfare of U.S. populations."[45] And the EPA gave prominent attention to the reciprocity argument, noting that "the only way to achieve an efficient allocation of resources for emissions reduction on a global basis—and so benefit the United States and its citizens and residents—is for all countries to consider estimates of global marginal damages."[46] It pointed to evidence suggesting that the reciprocity effect is real rather than theoretical—that is, that use of the global number by the United States actually does increase the likelihood that other nations will use that number.[47] Interestingly, the EPA downplayed the argument from moral cosmopolitanism; you have to squint hard to find it.

For a sense of what is happening on the ground, table 1.1 provides the numbers from the Biden administration in 2023 for carbon dioxide, methane, and nitrous oxide.[48]

We will get to the merits of the arguments in short order. But before doing that, it is important to see that they have played an important role in federal court.

Table 1.1
SC-GHG and near-term Ramsey discount rate

| | SC-$CO_2$ *(2020 dollars per metric ton of $CO_2$)* | | | SC-$CH_4$ *(2020 dollars per metric ton of $CH_4$)* | | | SC-$N_20$ *(2020 dollars per metric ton of SC-$N_2O$)* | | |
|---|---|---|---|---|---|---|---|---|---|
| Emission Year | Near-term rate | | | Near-term rate | | | Near-term rate | | |
| | 2.5% | 2.0% | 1.5% | 2.5% | 2.0% | 1.5% | 2.5% | 2.0% | 1.5% |
| 2020 | 120 | 190 | 340 | 1,300 | 1,600 | 2,300 | 35,000 | 54,000 | 87,000 |
| 2030 | 140 | 230 | 380 | 1,900 | 2,400 | 3,200 | 45,000 | 66,000 | 100,000 |
| 2040 | 170 | 270 | 430 | 2,700 | 3,300 | 4,200 | 55,000 | 79,000 | 120,000 |
| 2050 | 200 | 310 | 480 | 3,500 | 4,200 | 5,300 | 66,000 | 93,000 | 140,000 |
| 2060 | 230 | 350 | 530 | 4,300 | 5,100 | 6,300 | 76,000 | 110,000 | 150,000 |
| 2070 | 260 | 380 | 570 | 5,000 | 5,900 | 7,200 | 85,000 | 120,000 | 170,000 |
| 2080 | 280 | 410 | 600 | 5,800 | 6,800 | 8,200 | 95,000 | 130,000 | 180,000 |

*Source:* Table from https://www.epa.gov/system/files/documents/2023-12/epa_scghg_2023_report_final.pdf.

*Note:* Values of SC-$CO_2$, SC-$CH_4$, and SC-$N_2O$ are rounded to two significant figures. The annual unrounded estimates are available in article 5 in the appendix and at https://www.epa.gov/environmental-economics/scghg.

## DOMESTIC OR GLOBAL? LEGAL CHALLENGES

The social cost of carbon has been challenged in three cases. In *Louisiana v. Biden*, a district court enjoined agencies from adopting, employing, treating as binding, or relying on the work of the Interagency Working Group created by President Biden; from using the global cost of carbon; from failing to use discount rates of 3 percent and 7 percent; and from implementing Executive Order 13990.[49]

The court identified a range of concerns. First, it objected to the global rather than the domestic figure, apparently on the ground that use of the global figure exceeded the authority of federal agencies. Second, it said that any use of a social cost of carbon, ordered by the president and imposing "significant costs on the economy," would work a kind of transformation in federal regulatory law, and that any such transformation must be explicitly authorized by Congress under what is now called the "major questions doctrine."[50] Third, it objected to the use of a lower discount rate than the 3 percent and 7 percent figures in Circular A-4, on the ground that any shift from those figures would be arbitrary (see chapter 3). In fact, it offered a number of reasons to support its conclusion that the interim numbers from the Biden administration were arbitrary, including a failure to account for the factual findings made by the previous administration. (The court's decision did not survive appellate review, on the ground that those who sought to challenge the social cost of carbon lacked standing to do so at the time.)

In 2020, by contrast, a district court in *California v. Bernhardt* struck down a rule from the Bureau of Land Management (BLM), called the Waste Prevention Rule, whose content had been materially affected by the BLM's social cost of methane.[51] The court's analysis and conclusion focused above all on the agency's decision to use the domestic measure. In 2016, the BLM had relied on the analysis of the IWG in the Obama administration, which supported a conclusion that the total benefits of emissions reductions from the rule would range between $1.6 and $1.9 billion. Under the Trump administration, by contrast, the benefits numbers fell dramatically to a range between $66 and $259 million.

In generating the new numbers, the BLM relied on what it called an *interim domestic model*, focusing only on the benefits of emissions reductions to those living in the United States. The court held that the agency's decision was arbitrary and therefore unlawful. It ruled, first, that the agency could not defend its decision solely by reference to a relevant executive order. That order "did not and could not erase the scientific and economic facts that formed the basis for" the earlier estimate. No president, the court said, can "alter by fiat what constitutes the best available science." (Recall the earlier discussion of politics and technical expertise.)

For the Biden as well as the Trump administration, and for any future administration, this is an important conclusion: It suggests that if an agency is following an executive order, it is not immune from arbitrariness review. Rather, the substantive question remains, which is whether the agency is able to point, somewhere, to a rational justification for its choices—a question that bears on the legal validity of the use of the global or domestic number.

In the court's view, the mere fact that BLM is "the expert agency," not limited to peer-reviewed science, was not sufficient. First, "the social cost of methane is beyond BLM's expertise." Second, and more relevantly, the "interim domestic" model "is riddled with flaws." In offering this conclusion, the court referred to several of the arguments made by the IWG in 2016, above all the incompleteness argument. For example, the court said that the agency's estimate neglected the effects of greenhouse gas emissions on eight million U.S. citizens living abroad, and also on thousands of U.S. military personnel; on billions of dollars of physical assets owned by U.S. companies abroad; on U.S. companies affected by their trading partners and suppliers abroad; and on global migration and geopolitical security. The agency's failure to show "a rational connection between the best available science" and its estimate meant that its decision was arbitrary and therefore unlawful on the merits.

The court added that economists and scientists alike rejected the idea of focusing solely on the domestic effects. One reason for this rejection was the difficulty of offering an estimate of those effects in light of the limits of existing science—the epistemic argument. Another reason was the spillover effects, on the United States itself, of the international effects—the incompleteness argument.

In *Zero Zone v. Department of Energy*, a small business mounted a series of challenges to energy efficiency standards for refrigerator equipment.[52] One of the challenges involved the social cost of carbon. The plaintiffs urged, very broadly, that DOE was forbidden from considering environmental factors and, in the alternative, that DOE's analysis of the social cost of carbon was arbitrary and capricious.

The court did not engage these claims in detail. It referred only briefly to several objections, including the claim that the damage function was "determined in an arbitrary manner." The court stated that carbon could reasonably be deemed a global externality, that DOE had reasonably concluded that "national energy conservation has global effects," and that "those global effects are an appropriate consideration when looking at a national policy." The court rejected the various objections summarily, simply noting that the agency had responded to them in the rulemaking.

## IGNORANCE AND RELATIONSHIPS

What would it make sense for government agencies to do, as a matter of sound policy and morality, assuming that relevant law allows them to do what makes sense?

It is an understatement to use the word *challenging* to describe the task of calculating the social cost of carbon. Perhaps the central decision of the original IWG was to build on what were, at the time, the three leading integrated assessment models (IAMs), rather than to question them seriously or to attempt to make novel scientific judgments.[53] The three models are Dynamic Model of Integrated Climate and the Economy (DICE),[54] Climate Framework for Uncertainty, Negotiation, and Distribution (FUND),[55] and Policy Analysis of the Greenhouse Effect (PAGE).[56] These models attempt to specify the damage done by greenhouse gas emissions; they rely on both science and economics. If the goal is to monetize that damage, they provided (and continue to provide) a place to start.[57] The IAMs are evolving, and any particular account will rapidly go out of date. Nonetheless, the IAMs, both old and new, are controversial in terms of both science and economics; many people believe they depend on a great deal of guesswork.[58]

Exactly the same can be said of the modules on which the EPA relied in its draft report of 2022 and its final report of 2023. (Readers are encouraged to consult the final report for a valuable, detailed account of the underlying issues.[59]) Instead of drawing on the integrated assessment models, the EPA pointed to (1) a climate sensitivity module, (2) a damages module, (3) a discounting module (see chapter 3), and (4) a socioeconomics and emissions module (based on probabilistic projections for income, GHG emissions, and population). Because the four modules approach relies on the same kinds of projections that ground the integrated assessments models, we can take those models as illustrative of the challenges.

A number of years ago, Jonathan Masur and Eric Posner urged that "the three major economic models on which agencies rely are extraordinarily crude." In their view, "the cost of climate change will be high, but it is not clear how high, and one cannot conduct cost-benefit analysis of a regulation without knowing what its economic effect will be."[60] In a similar vein, and also a number of years ago, Robert Pindyck urged that the existing "models are so deeply flawed as to be close to useless as tools for policy analysis."[61] In his view, "the models' descriptions of the impact of climate change are completely ad hoc, with no theoretical or empirical foundation," and "the models can tell us nothing about the most important driver of the SCC, the possibility of a catastrophic climate outcome."[62]

Pindyck drew special attention to areas "where the uncertainties are greatest and our knowledge is weakest": climate sensitivity and the damage function.[63] With respect to climate sensitivity, Pindyck suggested that we know very little because "the physical mechanisms that determine climate sensitivity involve crucial feedback loops, and the parameter values that determine the strength (and even the sign) of those feedback loops are largely unknown, and for the foreseeable future may even be unknowable."[64]

With respect to the damage function, Pindyck urged that "we know almost nothing," which means that the developers of IAMs "can do little more than make up functional forms and corresponding parameter values."[65] Losses for individual regions, for agriculture, and for forestry are built on assumptions rather than data, and some of those assumptions are ad hoc.[66]

Hence, Pindyck's conclusion was that "the damage functions used in most IAMs are completely made up, with no theoretical and empirical foundation."[67] In any case, the IAMs do not say much about catastrophic outcomes, which is a quite serious gap.[68] Pindyck's conclusion was that we cannot reliably use the IAMs to establish a social cost of carbon, though we might explore plausible scenarios and make policy accordingly.[69]

Perhaps this conclusion is too pessimistic, and perhaps the newer modules approach, based on state-of-the-art evidence, is better.[70] Pindyck elaborated and updated his views in his 2022 book, *Climate Future*.[71] Even then, his conclusions there are broadly consistent with those just described. We do continue to learn (an understatement). But even if we know much more than Pindyck thinks, disaggregating the global harm, and producing some number to specify the damage done in the United States in particular, remains difficult. Contrary to Pindyck and Masur and Posner, we might believe that global assessments are (increasingly) helpful, and that science has progressed greatly there, while also insisting that it is not possible to project the harm from any ton of carbon emissions in the United States (or France or Germany or China). There is an argument that any specific numbers would be essentially a stab in the dark. If so, might it not be best to use a global number, which has a more plausible foundation in science and economics?

But this question should not be taken as rhetorical. The problem with the epistemic argument is that the domestic number is surely some fraction of the global number, and whatever the epistemic gaps justify, they do not justify use of that much larger number. Indeed, the IWG in the Obama administration acknowledged as much, suggesting that by one estimate (rooted in the integrated assessments models), the domestic number is about 7 to 10 percent of the global benefits, and that by another estimate (rooted in the U.S. share of global GDP), it is 23 percent of the global benefits.[72] The analysis by the Biden administration is consistent with the self-evident proposition that the domestic number must be significantly below the global number. For example, the number for the United States cannot possibly be four-fifths the global number, or three-quarters, or two-thirds.

In light of the epistemic challenges, perhaps agencies should use the latest research or some lower or upper bound, with some fraction of the global number reflecting the maximum damage to the United States, and another fraction representing the minimum. It might also be reasonable to specify some number between the upper or lower bounds, or perhaps to use the upper bound for reasons of prudence. For example, the executive branch might use the integrated assessment models or the modules approach and adopt some rules of thumb, working with U.S. population (about 4.25 percent of the world's total) and U.S. wealth (about 29.4 percent of the world's total). Perhaps the domestic number is reasonably estimated at 4.25 percent of the world's total, or 29.4 percent of the world's total, or somewhere in between.

Any judgment about which to use, or whether to use some weighted average, would of course be controversial. The only point is that taken by itself, the epistemic argument cannot possibly support use of the global number.

## INTERCONNECTEDNESS

In important respects, the interconnectedness argument is unambiguously right: if the social cost of carbon were limited to the direct harms imposed by climate change on U.S. citizens within the territorial boundaries of the United States, it would clearly be too low, even if we focused solely on the harms done by climate change to U.S. citizens. As the 2021 working group documented, more than nine million Americans live abroad, and they might well be adversely affected by climate change. At a minimum, the harms done to them should be included in a *domestic* cost of carbon.

Americans also have economic interests abroad; when those interests are harmed, Americans are harmed as well. Spelling out those interests would take a great deal of work, and it would show that a great deal is at risk. If we suppose that U.S. direct investment positions abroad are in the vicinity of

$6 trillion, we should be able to see immediately that a massive amount of U.S. wealth is vulnerable to the adverse effects of climate change. And what about residents of the United States who are not citizens? Should they not be counted as well, if we are speaking of that domestic cost?

As the 2021 working group also documented and as the 2022 EPA analysis elaborated, there is a more fundamental point. The harms done to foreign institutions and individuals will inevitably have spillover effects on U.S. citizens and U.S. interests. These effects must be counted, even if they are challenging and perhaps impossible to monetize. And if various nations suffer in economic or other terms as a result of climate change, those effects are likely to be manifested, in one or another way, in negative consequences for U.S. citizens and U.S. interests. Pressure on the immigration system is only one example. An adequate accounting of a "domestic" social cost of carbon must take those enormous consequences fully on board.

All of these claims are true, which means that it would be a mistake to adopt a social cost of carbon that is limited to the direct effects of climate change on Americans living within the territorial boundaries of the United States. If we are speaking of the domestic effects, we should use what might be called an *inclusive domestic number*, capturing the full set of effects on Americans. But what is that number?

Here again, there are serious gaps in knowledge. It would make sense to start by offering some accounting of Americans who live abroad and of the likely effects of climate change on them. It would also make sense to try to estimate the cascading effects of climate change on U.S. citizens and interests that result from the adverse effects elsewhere. That would be an exceptionally challenging task. Perhaps it would be unrealistic to generate point estimates. Here again, it might be possible to generate lower and upper bounds. Still, the central problem remains clear: However high the upper bound, it could not possibly be equivalent to the global number. It must be some fraction of it.

I now turn to the strongest arguments in favor of a global figure. Recall that the moral cosmopolitanism argument is that foreign lives matter, certainly insofar as U.S. actors are adversely affecting them. Recall too that the reciprocity argument is that because of the nature of the climate change problem, use of the global figure is essential or at least valuable for reasons of purely domestic self-interest.

## MORAL COSMOPOLITANISM

To fix ideas, assume that a pollutant—let us call it *pozone*—comes from a small number of power plants, all located in Detroit. Let us assume that pozone imposes nontrivial health risks. About half of the adverse health effects are expected to occur in the United States, and about half in Canada. Suppose that the EPA seeks to regulate stationary sources of pozone under some provision of the Clean Air Act.

### WHO COUNTS?

When the agency produces a regulatory impact analysis, which helps determine the stringency of its regulation, should it consider the harm to U.S. citizens only, or the harm to Canadians as well? Should it do so purely for reasons of transparency? Should it consider harms to Canadians in deciding on the right level of stringency? Are there legal constraints on its choices? An admittedly quite limited survey, described previously, suggests that agencies do not consider the harm to foreigners. Is that a mistake?

Let us begin with the easiest questions. As a matter of political morality, there is an overwhelmingly strong argument that the harm to Canadians ought to count.[73] If U.S. companies are causing deaths and illnesses in Canada, those are real deaths and illnesses: Why should they be ignored? Nor would it be arbitrary, as a matter of law, for the EPA to consider those harms, so long as the agency offered some kind of justification. The EPA might be committed to a kind of moral cosmopolitanism: *the lives of foreigners count.*

How much do they count? Monetary valuation raises independent questions. In recent years, federal agencies valued a statistical life at about $12 million, based on (American) willingness to pay to avoid statistical mortality risks. In poor nations, the value of a statistical life would of course be lower. I will turn to these issues in chapter 4. But for now, let us bracket questions of valuation and simply note that, according to moral cosmopolitanism, the value of a foreign life cannot possibly be $0.

As we have seen, moral cosmopolitanism raises many questions of both law and political theory, whether we are speaking of regulation, aid, military force, or anything else.[74] Many people believe that the job of political officials is to protect and help the people of their own nation, not the people of other nations.[75] Of course that belief would have to be justified, not simply asserted. But how? It is true that a politician who cared as much about noncitizens as citizens might not last in office very long. Imagine a candidate who proclaimed, "The world, not just us!" That candidate might not attract many voters. But we are speaking here of principle.

Within a utilitarian or welfarist framework, it might be speculated that if public officials embrace that belief, nations will actually be better off. A commitment to a form of nationalism might turn out to promote aggregate welfare: if U.S. officials focus on the United States, if French officials focus on France, if Chinese officials focus on China, if South African officials focus on South Africa, all of those nations might be better off than if all of those officials embraced moral cosmopolitanism. In that case, a form of nationalism might have a kind of rule-utilitarian defense, akin to the rule-utilitarian defense of the idea that mothers and fathers should focus on their own children, not on all children. There is, of course, a question whether that defense is convincing, and the answer must be largely empirical. Let us bracket that question here.

## BENEFITS AND HARMS

With respect to climate change, agencies might be more specific. They might insist that the kind of cosmopolitanism to which they are committed is narrow, modest, and constrained. They are *not* saying that the United States must engage in humanitarian intervention (and so use military force to save lives) or that the United States must give resources to foreigners (and so devote a large amount of the budget to foreign aid) or that the U.S. government should pay welfare benefits to poor people in other nations. Agencies should of course be open to the possibility that humanitarian intervention, development aid, and the like might rest, at least in part, on domestic self-interest. They might also be attuned to arguments from distributive justice (explored in more detail in chapter 3).

In the context of climate change, however, the claim is not about conferring benefits but avoiding harms—something like a tort. The claim is only that when U.S. companies and individuals impose harm on foreigners, that harm should be given full consideration when the United States is deciding on the stringency of its regulations. At the very least, the claim is that federal agencies should give consideration to that harm where governing law does not require otherwise.

The claim might be supported by reference to moral considerations. It might also be strategic: if the United States considers harm to Canadians, then Canada might consider harm to Americans as well. But that is a different point (and we will get there soon). Insofar as we are speaking of moral cosmopolitanism as I am characterizing it here, the basic claim is that the well-being of foreigners matters, certainly insofar as Americans are compromising that well-being. Taken in the context of climate change, the moral cosmopolitan argument thus specified is that if companies or others in the United States cause harm elsewhere, the United States has a moral obligation to consider that harm in deciding which regulations to impose. On what ground would a life of a person in Chad or Somalia, endangered by American emissions, count less than the life of a person in California or New York?

## NEGOTIATIONS

A possible counterargument would point toward the decisive importance of international negotiations. Let us stipulate that the United States seeks to ensure that every nation acts as morality requires. The problem is that if the United States chooses *unilaterally* to use the global number (say, $190), the outcome could be *worse*, on moral grounds, than it would be if it used the domestic number. The reason is that other nations might elect to free ride—and take advantage of new economic opportunities. If the United States uses the global number, for example, the price of fossil fuels might decrease, which means that other countries might be more rather than less enthusiastic about using them.

In this light, one might suggest that if we are speaking about the goals of moral cosmopolitanism, then a treaty of some kind, engaging the social cost of carbon in particular, must come first. Perhaps the United States ought to start with the domestic number and move to the global one *if and only if other nations do the same*. On this view, unilateral action is likely to be damaging to the very values for which moral cosmopolitanism purports to stand. In short: if help to foreigners is the goal, then choice of the domestic number is the place to start.

This argument is hardly preposterous, but it depends on speculative assumptions, and it is not a convincing response to the moral cosmopolitan argument. The simple claim behind that argument is that harms to foreigners matter. If that claim is correct, harms to foreigners must be counted in the social cost of carbon. It is true that if we were confident that counting those harms in that way would ultimately lead other nations to act less aggressively to reduce greenhouse gas emissions, we should hesitate. Those who emphasize moral considerations should always be prepared to consider the possibility that their morally preferred outcome, at time 1, might lead to outcomes at times 2, 3, and 4 that they would deplore. It is not impossible that use of the domestic number could be a strategic move toward the fulfillment of moral obligations.

There are no proofs here. But on reflection, it is far more likely, in this case, that the morally preferred approach at time 1 will lead to better, not worse, outcomes at times 2, 3, and 4—which brings us to the final argument.

## RECIPROCITY

Suppose that all nations used a domestic social cost of carbon, capturing the harm done by their own greenhouse gas emissions to their own citizens. If so, all nations would be losers. We can think of the climate change problem as a (repeated, wicked) prisoner's dilemma, in which individually rational actions by each nation produce losses for all. What is needed is a binding agreement by which all nations agree to scale back their emissions in a way that protects not only their own citizens, but those of every other nation as well. (The Paris Agreement can be seen as a good step in that direction, but it does not specify anything at all with respect to the social cost of carbon; see the appendix.) Here, in a nutshell, is the reciprocity argument: the use of the global figure by the United States might be regarded as a significant step toward ensuring widespread use of that figure.

In the end, this argument is convincing, but we need to be careful with it. It is hardly self-evident that the reciprocity argument justifies the use, by any one nation, of the global number. It is the very nature of a prisoner's dilemma that a unilateral action by one actor—in this case, the United States—does not lead to cooperation from other actors. Why should one nation impose costs on itself without receiving corresponding benefits? To be sure, moral cosmopolitanism provides something like an answer, but we are bracketing that answer here. If the use of a global number, by the United States, increased the likelihood that all of the world's nations would use that number to 100 percent, or very close to it, then we should be confident that the reciprocity argument is correct. But it cannot be said that the use of the global number, by the United States, essentially solves the prisoner's dilemma. Maybe other nations would simply say: Thanks! Good for you!

In 2010, the United States adopted a global number, and the other nations of the world did not promptly follow its lead. We did not see China, India, Brazil, and Russia suddenly adopting a global number and using it for regulatory or other purposes. (But importantly, some did do that; among others, Canada, France, Germany, Mexico, Norway, and the United Kingdom were influenced, and in some cases more than that.[76] I will return to that point.) Everyone should agree that the reciprocity argument would justify *use of the global number as part of an agreement by which many or all other nations agreed to do the same*. But recall once more that we are speaking of a prisoner's dilemma, which means that if the United States uses a global number, few other nations may elect to do so. Might China and India laugh all the way to the bank? How, exactly, would the prisoner's dilemma argument support a *unilateral* argument to use a global number when no general agreement to do that is in place?

The best answer, and the heart of the reciprocity argument in its most plausible form, is that with respect to climate change policy, the United States is an international leader (and of course it is hardly the only one). The United States can help establish a norm, and in that sense, it can incentivize other nations to do the same. The basic idea is that norms can solve prisoner's dilemmas, and they often do exactly that.[77] In the context of environmental harm, norms can do the work of law, and sometimes they substitute for law.[78] Indeed, they might ultimately form the foundation for law. The specific claims here are that (1) if the United States uses the domestic number, it might well be more likely that other nations will do that as well, to the detriment of all, and (2) if the United States uses the global number and thus leads by example, its role as an international leader might well lead toward a solution to this prisoner's dilemma. And indeed, existing evidence is consistent with this proposition; it suggests that strong action by the United States increases the likelihood that other nations will undertake strong action as well.[79]

There is no mathematical proof here, and reasonable people might make different predictions about likely scenarios.[80] As Jonathan Masur puts it, speaking of the relevance of the social cost of carbon to a treaty to control climate change:

> It may be that the United States has the greatest chance to convince China and India to enter into a climate change treaty if it behaves as if it values the lives of Chinese and Indian citizens at zero, as China and India may understand that they must agree to joint action if they wish to protect their own citizens. Or it might be that a treaty is most likely if the United States values Chinese and Indian lives equally to American lives, on the theory that they will demonstrate that the United States is acting in good faith. Or it might be that a treaty is most likely if the United States adopts some intermediate path.[81]

Fair enough. But if the United States uses the global number, it is fully plausible to think that it will become significantly more likely that other nations will, in the fullness of time, do so as well. Nor is this mere speculation; recall that a number of other nations have done exactly that. Indeed, the list of nations that have used a global social cost of carbon, or something like it, is very large and growing. It now includes Sweden, Switzerland, France, Finland, Denmark, Ireland, Slovenia, Costa Rica, South Korea, Iceland, South Africa, Chile, Portugal, New Zealand, Latvia, Mexico, Kazakhstan, and Estonia.[82] The European Union has used the global figure as well. To that extent, existing evidence is broadly consistent with the reciprocity argument. The key point is that use of the global number may well create an incentive for other nations to do the same, and thus reduce harms done to U.S. citizens and interests. If that is the goal, and especially in view of the sheer magnitude of the threat posed by climate change, use of the global number seems to be a more sensible bet than use of the domestic one.

To offer a little more detail: Suppose, as seems obvious, that use of the global number will not ensure that other nations will use that number. Suppose that it increases the likelihood that they will do so, and that it increases the likelihood that they will be willing to participate in an international agreement to reduce greenhouse gas emissions. Suppose that we do not know by how much it will increase those likelihoods. On one side of the ledger, use of the domestic number might reduce regulatory costs imposed on the American people. By how much? Probably not a great deal. Maybe not much at all. The answer depends on whether agencies decide to ramp up regulation significantly *by virtue of the use of the global number.* Use of that number will likely lead to increased costs—but not massively increased costs. On the other side of the ledger, use of the global number reduces an assortment of risks, genuinely catastrophic and otherwise, by some unknown and unknowable amount. In these circumstances, choice of the global number seems eminently reasonable, and almost certainly more reasonable than the alternative.

To be sure, there remain counterarguments. The point of the social cost of carbon is, of course, to capture the damage done by a ton of carbon emissions. Suppose that the appropriate global number is $200 and that as an upper bound, the appropriate domestic number is $50. By stipulation, the domestic damage is the latter, not the former. Even so, it would also be very good, and very important, for the United States to encourage other nations to reduce the damage they do to other nations, including the United States. Perhaps the United States should seek an international agreement to that effect. But on current assumptions, *the use of the global number is not a product of any such agreement.* (Recall that the Paris Agreement does not require it.)

For that reason (the objection continues), there appears to be a serious mismatch between (1) the goal of the basic enterprise, which is to specify the social cost of carbon, and (2) the reciprocity argument, which points to the need to produce certain kinds of action from other nations. The goal of the enterprise, of course, is to get the number right, not to negotiate with anyone. If the global number is the correct one to use because it will ultimately benefit citizens of the United States, the reason is that it will be helpful to negotiations that will produce that benefit, not that it captures the social cost of carbon.

If the objection is right, moral cosmopolitanism might seem to be a more convincing justification for use of the global number. The basic problem is that the global number does not, in any respect, reflect the actual damage done to the United States; it is a kind of bargaining chip, used at an early stage.

The objection does not lack logic on its own terms, but it is unpersuasive, so long as the reciprocity argument is described in the right way. In brief, the argument might take the following form: *We are choosing the global figure as a matter of policy. If every nation used the domestic figure, the citizens of all nations, including and most relevantly of the United States, would be worse off. Our choice of the global figure is part of a series of efforts to protect U.S. citizens and interests against that risk.* That argument is a compressed version of what was said in the Obama and Biden administrations, and it could easily be made without necessarily embracing moral cosmopolitanism. The reason is that it is founded on the importance of protecting U.S. citizens and interests, and it defends the use of the global number as an instrument toward ensuring that protection.

## AGENCY PRACTICE AND ADMINISTRATIVE LAW

The discussion thus far has focused on questions of basic principle, but it has implications for what happens on the ground, so to speak—for the practices of both agencies and courts.

For agencies, the most straightforward implication involves rulemaking. We have seen that under Executive Orders 12866 and 13563, agencies are required to quantify the costs and benefits of their regulations, and to proceed, to the extent permitted by law, only if the benefits justify the costs. I have argued that the global number is preferable to the domestic number, which means that quantification of the benefits of emissions reductions (or of the costs of emissions) should be based on the global number. Regulations involving motor vehicle emissions, power plant emissions, and much more should use the global number. But the arguments in favor of the global number extend to a much broader range of agency practices. Funding decisions should, for example, include that number, to the extent that analyses of costs and benefits are relevant. Congress should consider the global number in deciding what climate change programs to adopt and in deciding on their level of stringency. If and when the United States adopts a carbon tax, as it should, it ought to use that number.

More specifically, agencies should defend the global number, publicly and in court, principally by reference to arguments from moral cosmopolitanism and reciprocity, not by reference to the weaker arguments. Of the two stronger arguments, the reciprocity argument is the safer, because it focuses on domestic effects and sees the global number as a way of improving the welfare of American citizens; it does not make controversial claims about the appropriate weight to be given to the well-being of foreigners.

With respect to courts, we have seen that the governing statute is crucial, and also that certain canons of construction might turn out to be relevant. With respect to arbitrariness review, it is well established that agencies must give detailed justifications for their decisions.[83] The general implication is that for the social cost of carbon, a choice of the global number, or for that matter the domestic number, will be at serious risk if it is not explained in some way. As we have seen, the Trump administration's use of the domestic number was invalidated in part because it was not adequately explained. More specifically, use of the global number would be highly vulnerable if it rested solely or mostly on the epistemic argument, or on the incompleteness argument, including the admitted fact that American citizens and interests are at risk abroad. Indeed, a decision to justify the global number on either ground should and probably would be struck down.

At the same time, the moral cosmopolitanism argument should be sufficient to satisfy judicial review. The same is true of the reciprocity argument. And of the two, the latter is probably on firmer ground. The reason is that moral cosmopolitanism raises legitimate questions about whether and to what extent agencies can consider the costs and benefits of their regulations to foreigners. I have urged that they generally can do so, so long as statutes are ambiguous. Under arbitrariness review, there is no doubt that courts should defer to the executive's decision to rely on the reciprocity argument.

In light of the force of the two stronger arguments, it would be challenging to justify use of the domestic number in court. At a minimum, agencies would have to explain why they rejected the strongest arguments on behalf of the global number. With respect to moral cosmopolitanism, they might urge that the central obligation of a national government is to protect the citizens of that nation, not foreigners. With respect to reciprocity, they might argue that the goal of solving the prisoner's dilemma is best achieved by using the domestic number and by holding out the promise of the global number, or more dramatic cuts, as a bargaining chip in negotiations (recall Masur's concerns). Neither of these arguments is implausible in principle. If they are spelled out in some detail, the choice of the domestic number should not be struck down in court on arbitrariness grounds, even though it is inferior as a matter of policy.

## AN ALTERNATIVE APPROACH?

I have emphasized that specifying the damage from a ton of carbon emissions is an exceedingly challenging endeavor, and that some people believe that the specification is not possible.[84] If you are convinced by that argument, and nonetheless seek to assign monetary values, you might be tempted to seek another route. Because it is proving popular in prominent circles, let us say something about that route.

In an important and instructive essay, Nicholas Stern and Joseph Stiglitz urge that existing methodologies designed to identity the social cost of carbon, including those I have discussed, are fatally flawed.[85] Among other things, those methodologies focus on the externalities from greenhouse gases and ignore "other failures of fundamental importance," which the authors (reasonably!) claim are associated with

1. R&D and innovation;
2. capital markets;
3. networks (including grid structures, public transport, broadband, and recycling) in which there is extensive need for coordination, in which prices play only a limited role in that coordination, and from which a variety of positive externalities arise;
4. information (including around new products and the carbon content of those products); and
5. cobenefits (including air, water, and soil pollution).

In a discussion that bears directly on my focus here, Stern and Stiglitz also contend that it is crucially important to focus on moral issues, which include distributional questions (who is most at risk?) and the rights and interests of future generations. As they put it, "Climate change has very unequal impacts: it is usually the poorest people who are hit earliest and hardest; they live in more vulnerable areas, are less-well insured, and have weaker coping mechanisms. Those least responsible for emissions are among those most adversely affected." In their view, "common model choices may result in systematic bias, downplaying the importance of strong action on climate change and underestimating the social cost of carbon."

Importantly, Stern and Stiglitz also draw attention to the endogeneity of preferences.[86] What we prefer, and what we like, change over time. With new policies, our preferences might shift in desirable directions. For example, the costs of mitigation by behavioral adaptation might turn out to be lower than we anticipate. Suppose that people change their diets and end up eating less meat. If so, they might come to prefer those diets. In addition, an "increasing fraction of the population," they suggest, believes that the environment has intrinsic value and would put weight on it. Stern and

Stiglitz add that existing models devote too little attention to extreme risks and to uncertainty.

These various objections might well be taken one by one. Perhaps we could adjust existing approaches for some, most, or all of them. In fact, that would be an excellent idea. But instead of seeing if that is possible or desirable, Stern and Stiglitz suggest an altogether different course of action that does not depend on attempting to identify the social cost of carbon at all.

Their preferred approach would proceed in three steps: "First, describe the likely consequences from climate change, under current arrangements; second, examine how the economy and emissions could be managed to give a good chance of stabilizing at different temperatures; and third, combine these two elements into a judgement on an approach to a temperature target." In essence, they urge, we might adopt a constraint such that the temperature (properly modeled) never increases beyond 2°C (compared to preindustrial levels). With this constraint in place, we can try to calculate the social cost of carbon along a path where temperature increases are constrained below 2°C. In that way, we could produce a social cost of carbon, *not* by projecting the damage from a ton of carbon emissions, but by specifying what the price on carbon would have to be to ensure that the increase in global temperature does not exceed 2°C.

Let us suppose the resulting number would be one hundred dollars. If the goal is to specify the social cost of carbon, would that be reasonable? At first glance, it would not be. In fact, it seems to make no sense at all. To figure out what target we should pick, we need to know the costs and benefits of various targets; the target must be a *product* of that assessment, and it should not be the basis of an assessment of the social cost of carbon. But to say the least, Stern and Stiglitz know what they are doing, so let us identify the assumptions on which use of the target might make sense.

Suppose that an analysis of all relevant costs and benefits suggested that a global cap of an increase of 2°C would in fact be optimal, or in other words that a cap of 1.99°C would be too low and that a cap of 2.01°C would be too high. If so, the global cap of 2°C is exactly the right number. In that case, it follows that the social cost of carbon that one would emerge from that cap would also be the right number—and it would be *identical* to the social cost of carbon that would emerge from the right IAM. It follows that (1) *if* there were a world government, (2) *if* the world government could reasonably conclude on the basis of an analysis of all relevant benefits and costs that the maximum acceptable increase in warming would be 2°C, and (3) *if* the world government could produce and make binding a monetary figure that would reflect what was necessary to ensure point 2, then the Stern and Stiglitz argument would be entirely convincing. And if there were no world government, but if the nations of the world could agree to a binding treaty, the same conclusion would follow: the Stern and Stiglitz argument would be entirely convincing.

Turn now to a puzzle. In light of all the uncertainties, is the maximum acceptable increase in warming 2°C? Or 1.99°C? Or 1.987°C? Or 1.875°C? Or 1.752°C? Any claim that it is 2°C, or 1.5°C, should perhaps not be taken as a purely scientific and economic conclusion, if only because such numbers are suspiciously round!

But let us put these questions to one side and assume that, for one or another reason, the maximum acceptable increase is warming 2°C (or 1.5°C, or something identifiable). Even if that is so, there is no world government, and there is no global treaty of the relevant kind, which means that no institution has the authority to insist on a monetary figure that is necessary to ensure a maximum increase in warming of 2°C.

So the real question is this. Suppose that acting on its own, the United States chooses that monetary figure—say, one hundred dollars—and calls it the *social cost of carbon*. Would that be a good idea?

Maybe so. Maybe it would be a laudable effort to exercise global leadership. But the specified amount would not reflect the social cost of carbon at all. It would not reflect the amount of damage done by a ton of carbon emissions. It would instead reflect the tax that would ensure that the world would not exceed the specified limit, if the world's nations agreed to that tax. That is not a social cost of carbon.

## A PLEA

Development of a social cost of carbon raises numerous challenges. A central question is whether to choose the domestic or instead the global figure. It is true that specialists find it exceedingly difficult to generate a domestic number, but by itself, that difficulty is not a sufficient justification for using the global number—which must, in any account, be significantly higher than the domestic one. It is also true that climate change imposes risks to U.S. citizens and U.S. interests that go far beyond the risks associated with higher temperatures within the territorial boundaries of the United States. For that reason, any domestic figure must be inclusive of the full set of (domestic) harms. Calculating the correct domestic figure, in that light, presents even more serious challenges, but again, there is a difference between an inclusive domestic figure and the global figure, and the former is inevitably a fraction of the latter.

On moral grounds, there is a strong ethical argument in favor of a narrow and constrained form of cosmopolitanism: when companies and individuals in the United States impose harms on foreigners, those harms are real, and they should be counted. There is also a strong argument in favor of using the global figure as part of an assortment of strategies designed to ensure an effective response to the problem of climate change, one that protects U.S. citizens and their interests against the harms coming from emissions from other nations. This is the argument from reciprocity. Even though it rests on a degree of speculation, the speculation is more than plausible. In this light, climate change cosmopolitanism, whether based on moral or strategic grounds, is the right approach to the social cost of carbon.

# 2 RICH NATIONS, POOR NATIONS

Imagine that Canada discovered a new source of energy. Use of the new source would be an economic boon to Canada, and it would impose no environmental harm on that nation. At the same time, the new energy source would make the air much dirtier in the Central African Republic. (If that seems implausibly unrealistic, recall that we are using our imagination here.) The citizens of the Central African Republic would be entitled to object that Canada is enriching itself at their expense. Isn't the climate change problem exactly like that? My ultimate answer is no: It is not exactly like that. But it is more than a little like that. It is enough like that—which is my central conclusion in this chapter.

These issues are at the heart of international negotiations with respect to climate change. The intensely disputed idea of loss and damage has been central to those negotiations, with poor nations, especially vulnerable to climate-related risks, asking for a loss-and-damage fund to compensate them for relevant harms.[1] In 2022, the world's nations finally agreed to establish such a fund.[2] Sherry Rehman, Pakistan's climate minister, spoke for many in saying: "This is how a 30-year-old journey of ours has finally, we hope, found fruition today."[3]

To be sure, the idea of a loss-and-damage fund raises many questions—scientific, political, and ethical.[4] How big should the fund be? Who contributes to it, and in what proportions? Who exactly gets the money? Where shall the fund be housed, and who should be responsible for managing it? These questions are complex and intensely debated, and the difficulties of achieving consensus on any one of them became immediately evident once participating nations began negotiating over how the fund would work.[5] With what framework might they be answered? In important circles, that question continues to hang in the air.

Begin with some simple facts. The United States long led the world in annual greenhouse gas emissions, but China has greatly surpassed the United States. Any particular numbers will rapidly become out of date, but the two leading emitters now account for over two-fifths of the world's annual emissions.[6] The emissions of the United States and China promise to impose serious losses on other nations and regions, emphatically including Europe but perhaps above all India and Africa. For this reason, it is tempting to argue that both nations are, in a sense, engaging in wrongful or tortious acts against those nations that are most vulnerable to climate change. As we have seen, the top emitting nations have been imposing serious harms on other nations, and poor nations are the disproportionate losers.

This argument might seem to have special force as applied to the actions of the United States. While the emissions of the United States are growing relatively slowly, that nation remains the largest contributor to the existing "stock" of greenhouse gases.[7] Because of its past contributions, does the United States owe compensation to those nations, or those citizens, who have been harmed or who are most likely to be harmed by climate change? Principles of corrective justice might seem to require that the largest emitting nation pay damages to those who are hurt. That is the basic idea behind the notion of loss and damage, and in particular of a loss-and-damage fund.

Apart from that idea, the United States has the highest gross domestic product of any nation in the world, and its wealth might suggest that it has a special duty to help to reduce the damage associated with climate change. This is a point about distributive justice. Are the obligations of China, the leading annual emitter, equivalent to those of the richer United States, the second-leading annual emitter? Does it not matter that China's per capita emissions rate remains significantly below that of the United States?[8]

We should emphasize here that while China is significantly poorer than the United States, it is not among the poorest nations in the world. Any particular account will change over time, but of the world's 193 nations, the poorest include Burundi, Somalia, Mozambique, the Central African Republic, Madagascar, Sierra Leone, Democratic Republic of the Congo,

Niger, Malawi, and Liberia. China is much wealthier. To categorize it as poor is far too simple, not least in light of its rapid growth over the last decades. This point shows that we do not have a simple division between rich nations and poor nations, and also that if we are speaking of poor nations, it would be a mistake to group China with them. For the sake of exposition, I will often speak of rich nations and poor nations, but the use of those adjectives should not obscure the fact that we are speaking of a continuum, not a dichotomy.

With that qualification in mind: As a matter of principle, and putting climate change and corrective justice to one side, there is a good argument that in many domains, resources should be redistributed from the richest nations and richest people to the poorest nations and poorest people. As we have seen, high levels of foreign aid might be justified on grounds of domestic self-interest, but they also reflect, in part, a moral judgment to that effect. (Are current levels high enough? Too high? Should we worry about the risk that the resources will be misused or wasted by incompetent or corrupt governments? How much should we worry about that? Let us bracket these topics, important though they are.) It is also true that if rich people lose a certain amount of money, and poor people receive that amount of money, total welfare will be increased. A debt of $100,000 is nearly meaningless to a billionaire, but it is a catastrophe for someone at the edge of subsistence. If rich people in rich countries lost a specified amount of money, and if poor people in poor countries received that same amount of money, total welfare would be much higher.

Apart from the overall welfare gain, those who focus on the relief of deprivation and on the importance of helping those at the bottom (prioritarianism, on which more to come) would suggest that a level of redistribution is an excellent idea and perhaps required as a matter of justice. Let us sharpen this point by supposing that poor countries imposed climate-related risks on rich countries—that (say) Somalia, the Central African Republic, Burundi, and South Sudan engaged in activity that endangered people in the United States and Canada. If so, the United States and Canada would have a strong corrective justice argument. But they would not, of course, have an argument for distributive justice. In the context of climate change, the two arguments tend to march hand-in-hand.

We will encounter these claims at various stages. As noted, national boundaries certainly matter (if only for reasons of feasibility), and apart from what is feasible, it might be true that insisting on their relevance imposes legitimate constraints on redistribution from wealthy countries to poor countries. But let us simply stipulate that the justification for some such redistribution is compelling. How does this bear on the ethical issues?

## OPTIONS

The Paris Agreement was adopted in 2016 (see the appendix). It suggests the existence of a strong international consensus in support of the view that the nations of the world would benefit from significant steps to control greenhouse gas emissions. If all of the major emitting nations firmly agreed to such steps, the benefits would almost certainly exceed the costs. To be sure, there is a lot of ambiguity in the term *significant steps*, and there are of course serious disagreements about the appropriate timing and severity of emissions reductions. The social cost of carbon can be seen as a way of mediating those disagreements. And while a great deal has happened since 2016 (to say the least), the Paris Agreement continues to provide the foundation for international discussions.

The Paris Agreement allows a great deal of flexibility to nations, building as it does on the idea of *nationally determined contributions*. Those contributions come from national plans, not from an international agreement itself, and even so, they are nonbinding. They must be updated every five years. The overall goal is to attempt to limit total warming to 1.5°C (compared to preindustrial levels) and to aim to keep warming well below 2.0°C.

In light of the nature and magnitude of the climate change problem, this is a much softer and more flexible approach than would be ideal. It is obviously a product of the complex motivations of national leaders, who have not yet been willing to support a firmer and more aggressive approach. With some combination of determination and luck, the Paris Agreement might in the fullness of time provide the functional equivalent of that approach, or the foundation for it. The best bet, and the most optimistic, is that incremental improvements in the agreement, and multiple actions and decisions with respect to nationally determined contributions, will produce large emissions reductions. This is not a book about current events, but as of 2025, the jury is out.

It should be clear that the Paris Agreement takes an unusual approach, produced by political necessities. In policy circles, there has long been an understanding that if the nations of the world do undertake an effort to reduce greenhouse gas emissions, they should select one of two possible approaches.[9] The first is an emissions tax, designed to capture the externalities associated with greenhouse gas emissions. With an approach of this kind, the tax might be uniform, capturing those externalities. Citizens of Russia, China, India, the United States, France, and so forth would all pay the same tax. There is a disagreement about the proper magnitude of the tax, and here again, the social cost of carbon might provide the right place to start (and possibly to end). At the present time, a global carbon tax of one hundred dollars would seem to be a reasonable starting point.

The second approach would involve a system of cap-and-trade, akin to that in the now-obsolete Kyoto Protocol (adopted in 1997 and entering into force in 2005). Under such a system, nations might create a worldwide cap on aggregate emissions. A cap-and-trade system would require judgments about (1) the appropriate cap and (2) the initial allocation of emissions rights. Both of these judgments are hard. In one version, roughly embodied in the Kyoto Protocol, *existing emissions levels* would provide the foundation for initial allocations; nations would have to reduce by a certain percentage from those existing levels. But the use of existing levels is highly controversial. Why should nations be entitled to start there? What if their existing levels are very high? One answer might be that the percentage reductions would be especially high, and so especially costly, for those with high current levels.

Analytically, a cap-and-trade system is not very different from a uniform carbon tax, which does not impose a global cap, but which would ensure a reduction in emissions, whose size would depend on the magnitude of the tax. There are active debates about which is better; the current consensus is that a carbon tax would have significant advantages.[10] To be sure, both face serious problems of feasibility, as shown by the Paris Agreement, which (it will be recalled) adopts instead the idea of nationally determined contributions.

We could easily imagine a host of other approaches. Nations might impose, and nations have imposed, regulations covering fuel economy, energy efficiency, and emissions from power plants. Nations might give, and have given, subsidies designed to promote electric vehicles and use of energy sources that emit low levels of greenhouse gases (or perhaps no greenhouse gases at all). It is generally agreed that regulations are less efficient than carbon taxes and cap-and-trade. They can produce significant emissions reductions, of course, but at a much higher cost. Subsidies force taxpayers to pay for emissions reductions, and to that extent have disadvantages compared to carbon taxes and cap-and-trade. With carbon taxes, nations get revenue; with subsidies, nations lose revenue. If deficits are important, taxes look a lot better than subsidies. And with carbon taxes, polluters pay, not taxpayers.

There is another problem. With subsidies, governments must also decide *whom to subsidize*, and they might make the wrong bets. A tax on greenhouse gas emissions has comparative neutrality; companies and individuals can figure out, on their own, what to do or buy instead. Some people much like the idea of industrial policy; they think that governments are in a unique position to give a boost to (say) solar or wind, or to press in the direction of electric cars. Enthusiasm for industrial policy is certainly higher than it was twenty years ago. Maybe the enthusiasts are right, and if so, subsidies have advantages. But if the goal is to let markets pick winners, rather than public officials, then carbon taxes are a lot better. How can public officials know what clean energy source is best? Now? Twenty years from now? It is also true that regulations and subsidies also have their own distributional effects; they will create different winners and losers from (say) a carbon tax.

Let us bracket these issues and emphasize that an agreement to control greenhouse gas emissions loses nearly all of its point if only a few nations are willing to participate. Fortunately, the Paris Agreement does have broad participation, notwithstanding the vagueness of the idea of nationally determined contributions. To see why that is so important, consider the old Kyoto Protocol, which required most of the industrialized world to cut emissions significantly. Because developing nations refused to accept any emissions restrictions, a prominent study offered this stunning finding: full compliance with the agreement would have reduced anticipated warming by merely 0.03°C by 2100.[11]

Projections are frequently changing, but not so long ago, the Intergovernmental Panel on Climate Change provided a "best estimate" warming of 1.4°C to 4.4°C by 2100, under a "business as usual" scenario.[12] Relatively little would be gained if all nations complied with obligations of this kind and reduced those figures to a range of 1.37°C to 4.37°C.[13] A more optimistic estimate found that the Kyoto Protocol might have reduced global warming by as much as 0.28°C by 2100, and the difference between business as usual warming and warming between 1.52°C and 3.72°C is not exactly trivial.[14] But if developing nations were included, far more significant reductions could be anticipated (as we are about to see). The need for broad participation has important implications for questions of efficiency, effectiveness, and justice.

## EMITTERS

To understand the ethical issues and the motivations of the various actors, it is important to appreciate the disparities in emissions across nations. We do not have clear data on the costs of emissions reductions for different nations, but it seems reasonable to begin with the assumption that on average, the largest carbon emitters would bear the largest burdens from (say) a worldwide carbon tax, and the small emitters would suffer the least.[15] (As Bob Dylan once sang, "When you ain't got nothin', you got nothin' to lose.") For a snapshot, consider table 2.1.

As early as 2004, the United States and China emerged as the top emitters, and they now account for over 45 percent of the world's total, with China itself approaching one-third. (Please pause over that; it is staggering.) If the goal is to understand the ethical issues and the costs of controls,

Table 2.1
Share of global emissions, 2020

| | 2020 |
|---|---|
| United States | 13.3% |
| OECD Europe | 10.4% |
| China | 31.8% |
| India | 6.2% |
| Japan | 3.2% |
| Africa | 3.6% |
| Russia | 5.4% |

*Source:* United States Energy Information Administration, Table A10: World carbon dioxide emissions by region, reference case, in *International Energy Outlook 2021* (Oct. 2021), https://www.eia.gov/outlooks/ieo/data/pdf/ref/A10_r.pdf.

however, this table does not tell us nearly enough; we need to know future projections as well. Existing projections suggest that the largest contributors are likely to continue to qualify as such—but that major shifts will occur, above all with emissions growth in China and India, and emissions reductions in Russia. For a sense of trends, begin with changes between 1990 and 2017, shown in table 2.2.

With a high degree of humility, we can project changes to 2050. At that time, non-OECD countries are expected to contribute no less than 72 percent of total emissions, with 28 percent coming from OECD countries.[16] At that time, the United States is expected to be far below China. This projection is startling insofar as it suggests that India will surpass the United States as early as 2040, and that China and India will contribute nearly 40 percent of the world's total by 2050. See table 2.3.

Table 2.2
Carbon dioxide emissions changes, 1990–2017 (emissions of $CO_2$ from energy-related sources only)

| | 1990–2017 |
|---|---|
| China | 343.2% |
| United States | –0.9% |
| India | 308.6% |
| South Korea | 158.9% |
| Iran | 231.2% |
| Indonesia | 269.8% |
| Saudi Arabia | 252.2% |
| Brazil | 131.8% |
| Spain | 25.1% |
| Pakistan | 227.7% |
| Poland | –11.3% |
| EU-28 | –20.3% |
| Germany | –23.5% |
| Ukraine | –75.1% |
| Russia | –29.0% |

*Source:* Emissions of $CO_2$ from energy-related sources only. See International Energy Agency, *CO2 Emissions from Fuel Combustion* II.4–II.6 (2019).

Table 2.3
Relative contributions of annual carbon dioxide emissions by country/region (approximate % of worldwide emissions)

| | 2020 | 2025 | 2030 | 2040 | 2050 |
|---|---|---|---|---|---|
| United States | 13.3% | 12.5% | 12.2% | 11.4% | 11.2% |
| OECD Europe | 10.5% | 10.1% | 9.6% | 8.9% | 8.6% |
| China | 31.8% | 30.1% | 28.8% | 26.5% | 24.5% |
| India | 6.2% | 7.8% | 9.0% | 11.9% | 13.6% |
| Japan | 3.2% | 3.0% | 2.8% | 2.5% | 2.0% |
| Africa | 3.6% | 3.9% | 4.1% | 4.4% | 4.8% |

*Source:* Emissions of $CO_2$ from energy-related sources only. See International Energy Agency, *CO2 Emissions from Fuel Combustion* II.4–II.6 (2019).

The numbers thus far refer to *flows*: how much a given nation emits on an annual basis. Of at least equal relevance for ethical issues are *stocks*: how much a given nation has, over time, contributed to the current stock of greenhouse gases in the atmosphere. Tables 2.4 and 2.5 tell the story.

Note that the United States accounts for more than one-fifth of the total, but that China is not terribly far behind (and that the gap is closing every day). Note too that taken together, Russia, Japan, India, Germany, and the United Kingdom well exceed the United States.

It is important to emphasize that greenhouse gases, and especially carbon dioxide, tend to dissipate slowly, which helps explain why countries that industrialized earlier have contributed more to the stock than countries that industrialized later, even though the latter might today contribute more on an annual basis. About half the $CO_2$ emitted in 1907 still remains in the atmosphere. And if the world stopped emitting $CO_2$ today, the stock of $CO_2$ in the atmosphere in 2107 would remain at about 90 percent of what it is now.[17]

This point greatly matters to many issues; it helps to explain, for example, why even significant emissions reductions will reduce but hardly halt anticipated warming. Unilateral action, even by the largest emitters, cannot affect the existing stock, and by definition, it will do nothing (directly) about the rest of the flow.

Table 2.4

Cumulative emissions (1850–2021)

| | $CO_2$ |
|---|---|
| United States | 558469 |
| China | 365721 |
| European Union | 403614 |
| Russia | 174664 |
| Japan | 77448 |
| India | 118658 |
| Germany | 115196 |
| United Kingdom | 98133 |
| Canada | 46045 |
| South Korea | 23552 |

*Source:* See *Historical GHG Emissions*, ClimateWatch, https://www.climatewatchdata.org/ghg-emissions?breakBy=countries&calculation=CUMULATIVE&end_year=2021&gases=kyotoghg®ions=WORLD&source=PIK&start_year=1850. $CO_2$ is in megatons.

Table 2.5

Cumulative emissions (1850–2021)

| | $CO_2$ Share |
|---|---|
| United States | 22% |
| China | 14% |
| European Union | 15.9% |
| Russia | 6.7% |
| Japan | 3.0% |
| India | 4.6% |
| Germany | 4.4% |
| United Kingdom | 3.8% |
| Canada | 1.76% |
| South Korea | 0.90% |

*Source:* See *Historical GHG Emissions*, ClimateWatch, https://www.climatewatchdata.org/ghg-emissions?breakBy=countries&calculation=CUMULATIVE&end_year=2021&gases=kyotoghg®ions=WORLD&source=PIK&start_year=1850. $CO_2$ is in megatons.

## VICTIMS

Which nations are expected to suffer most from climate change? Of course, the precise figures are greatly disputed; as we have seen, the extent of the damage in 2100 in particular nations cannot be specified now, which is why it is so difficult to identify a domestic social cost of carbon.[18] But it is generally agreed that the poorest nations will be the biggest losers by far.[19] The wealthy nations, including the United States, are in a much better position for three independent reasons.[20] First, a smaller percentage of their economy depends on agriculture, a sector that is highly vulnerable to climate change. Second, the wealthy nations are generally in the cooler, higher latitudes, which also decreases their vulnerability.[21] Third, they have much more in the way of adaptive capacity.

To get a handle on the problem, let us assume, somewhat randomly, that warming will be 2.6°C, and consider, just for illustration and with many grains of salt, a 2021 set of estimates of how the harms are likely to vary across nations and regions (see table 2.6).

By these estimates, it is readily apparent that some nations are far more vulnerable than others.[22] India and China are anticipated to be massive

Table 2.6
Damages of a 2.6°C warming as a percentage of GDP

| | |
|---|---|
| India | 13.9 |
| Japan | 4.5 |
| United States | 3.9 |
| China | 9.2 |
| Russia | 5.8 |
| South Africa | 9.2 |

*Source:* Data from the Swiss Re Institute report in April 2021 using the *high stress* model, which provides relatively conservative estimates. See *The Economics of Climate Change*, Swiss Re Inst. (Apr. 22, 2021), https://www.swissre.com/institute/research/topics-and-risk-dialogues/climate-and-natural-catastrophe-risk/expertise-publication-economics-of-climate-change.html.

*Note:* Swiss Re is "one of the world's largest providers of insurance to other insurance companies." Christopher Flavelle, *Climate Change Could Cut World Economy by $23 Trillion in 2050, Insurance Giant Warns*, N.Y. Times (Apr. 22, 2021), https://www.nytimes.com/2021/04/22/climate/climate-change-economy.html.

losers. India is expected to experience devastating losses in terms of both health and agriculture. About 80 percent of people in India live in areas deemed highly vulnerable to the extreme weather caused by climate change.[23] While many Indians already live in what are called *hotspots* for the negative effects of climate change, in 2050 more than 148 million Indians will reside in *severe hotspot* regions.[24] For Africa, the major problem involves health, with a massive anticipated increase in climate-related diseases and deaths. By 2030, it has been estimated that the effects of climate change could raise death rates on the continent by 60 to 80 percent.[25] Perhaps that number is far too high. But without a great deal in the way of adaptation, significant increases in deaths will be driven largely by malaria and diarrhea.[26]

To be sure, the United States also faces significant threats to both agriculture and health. Consider just one study of the long-run effects of climate change on a range of economic variables in the United States.[27] The study offers both optimistic projections, including a high level of adaptation and low warming, and pessimistic projections, involving little adaptation and high warming. For 3.2°C warming, the most optimistic case projects a decrease of 1.2 percent in GDP. The most pessimistic case projects losses of 9.2 percent of GDP at 3.2°C. In view of the many sources of uncertainty, these estimates should hardly be taken as authoritative. On the one hand, the most optimistic case might turn out to be too pessimistic; on the other hand, the risk of catastrophe greatly complicates matters.[28] But it is clear that that the greatest vulnerabilities are not in the United States.

In light of all this, it seems useful to analyze the ethical issues by assuming that the world would greatly benefit from a treaty, firmer than or building on the Paris Agreement, to control greenhouse gas emissions; that some nations would have to pay far more than others to reduce their emissions; and that some nations are far more vulnerable to climate change than others. Who pays for emission reductions? Who pays for the damages? There is also the question of adaptation. Who pays for that?

## DISTRIBUTIVE JUSTICE

Should the United States give resources to poor countries to help them to deal with climate change? Would that be desirable from the standpoint of distributive justice? The simplest answer to these two questions is "yes."

If poor nations are in need and if they face serious risks, wealthy nations should help them. Because a given amount of money is worth more in the hands of the poor than the rich, such help will increase welfare. It is also supported by considerations of distributive justice. A rich person should help a poor person; a rich country should help a poor country (assuming that the help will actually be provided to those who need it). But even with that answer, there are two problems.

*First:* If the goal is to help poor people in poor nations, why not transfer cash, rather than more targeted assistance? What is the point of prescribing how money must be used? Why should the use of the money be specified (for, say, green energy or adaptation to climate-related risks)? Wouldn't it be better to allow poor nations to make the relevant choices for themselves? This is, of course, a pervasive problem in development policy. In general, cash is indeed best. Still, it is possible that targeted assistance, focusing on mitigation or adaptation, will produce larger gains than cash grants, perhaps because of the risk of corruption, perhaps because the cash will not be used wisely or well. There is a large literature on this issue; I simply flag the question here.[29]

*Second:* Nations are not people. They are collections of people, ranging from very rich to very poor. Wealthy countries, such as the United States, have many poor people, and poor countries, such as India, have many rich people. Still, the median member of wealthy nations is wealthier than the median member of poor nations, which makes it plausible to think that if wealthy nations help poor ones, the distributive effects are likely to be good. For example, the Americans who might be asked to make the relevant payments are, on average, wealthier than the Indians who have the ability to pay less. It remains true that asking the United States, as such, to make payments to a poor nation, as such, is hardly the best way of achieving the goal of transferring wealth from the rich to the poor.

Nothing said here denies the proposition that if the United States is willing to give money to poor nations to help them to handle climate-related harms, poor people are likely to be benefited on balance. But if the goal is to transfer resources from wealthy people to poor people, alternative policies might be better. It should be clear that these claims apply broadly to efforts to invoke distributive justice to ask wealthy nations to participate in international agreements from which other nations might gain. Putting claims of corrective justice to one side, redistribution from wealthy nations to poor nations might be a matter of rough justice—a point to which I will return. And as we have seen, such redistribution might also be in the self-interest of the donors, because it is in one's interest to have friends.

How much redistribution? It would be foolish to specify any particular amount. Let us just say: a lot.

## CORRECTIVE JUSTICE

Corrective justice arguments are backward-looking, focused on wrongful behavior that occurred in the past. Corrective justice therefore requires us to look at stocks rather than flows. We have seen that even though China is now the world's leading greenhouse gas emitter, the United States has been the largest emitter historically and continues to bear the greatest responsibility for the stock of greenhouse gases in the atmosphere.

Of course, a disproportionate share of the stock of greenhouse gases can be attributed to other long-industrialized countries as well, such as Germany and Japan, and so what is said here about the United States can be applied, mutatis mutandis, to those other countries. The emphasis on the United States is warranted by the fact that the United States has contributed more to the existing stock than any other nation. Indeed, what is said here about the United States also applies to China, even though it is not (yet) rich.

In the context of climate change, the corrective justice argument is simply that the United States has wrongfully harmed the rest of the world—especially warmer nations, low-lying nations, and other nations that are most vulnerable to climate change—by emitting greenhouse gases in vast quantities. On a widespread view, corrective justice requires that the United States devote significant resources to remedying the problem, perhaps by paying damages. Recall the loss-and-damage fund. Pakistan, for example, might be taken to have a moral claim against the United States, one derived from the principles of corrective justice, and if so, the United States has an obligation to provide a compensatory remedy to Pakistan. Because Pakistan is especially vulnerable to climate change, that nation can be taken as a placeholder for those at particular risk.

This argument enjoys a great deal of support in prominent circles and seems intuitively correct.[30] Many people in poor nations insist on it. In the end, I will suggest that the argument is correct enough. The apparent simplicity of the argument, however, masks difficulties, which distinguish the case of current and future emissions (the topic of chapter 1) from that of emissions in the past.

## THE AGGREGATION PROBLEM

The United States is not a person, nor is Pakistan. Corrective justice is typically a matter between individuals, not entities. To see the problem, consider the International Court of Justice suit brought by Bosnia and Herzegovina against Serbia, charging Serbia with genocide during the Yugoslav civil war of the early 1990s.[31] Suppose that Bosnia and Herzegovina had won this case and Serbia had been forced to pay reparations. No such entity called *Serbia* can pay out of its pocket; the reparations would be financed out of general revenues, paid for by taxes. Thus, the effect of the remedy would be to raise taxes or reduce government services for some or all Serbians, while benefiting—in the form of lower taxes, lump-sum payments, or increased government services—some or all Bosnians.

Can such an effect be justified? Very possibly, but the point for present purposes is just that talk of corrective justice between states can be only a metaphor. States do not act; individuals act. States do not have mental states; individuals have mental states. To evaluate the moral considerations touching on claims between states, one needs to penetrate the veil of the state and consider the activities of the people who operate their governments and the people who are affected by their policies. Indeed, Bosnia's claim was not based on any injury to "Bosnia"; it was based on an injury to Bosnians. If any state as such was a victim of the civil war, it was Yugoslavia, which was broken into pieces, and yet no one thinks that "Yugoslavia" or some representative or successor has a claim against whoever was responsible for its dismemberment.

From the perspective of corrective justice, these points are concerning. Wrongdoers should compensate victims for their losses, and yet the crude state-to-state remediation scheme results in innocents being punished and nonvictims being compensated. If this point is not immediately intuitive, it is because states tend, wrongly, to be personified. To evaluate any claim against the United States (or other nations) for wrongfully causing climate change, we must consider the actions of individuals (including corporations), and the effects on individuals, and try to avoid referring to states qua states.

## WHO'S THE WRONGDOER?

The current stock of greenhouse gases in the atmosphere is due in significant part to the behavior of people living in the past. Of course it is also true that older people, as a class, have contributed a significant share, but much of the stock is due to the acts and omissions of people who are dead. The basic problem for corrective justice is that dead wrongdoers cannot be punished or held responsible for their behavior or forced to compensate those they have harmed. Holding Americans today responsible for the activities of their ancestors is not obviously fair or reasonable on corrective justice grounds, at least not unless contemporary Americans can be said to have benefited from the actions of their ancestors (an issue to be discussed shortly).

Emphasizing that the principal beneficiaries of greenhouse gas reductions will be future generations, Thomas Schelling argues that "greenhouse gas abatement is a foreign aid program, not a saving-investment problem of the familiar kind."[32] We should qualify Schelling's point: perhaps what is sought is resources for loss and damage or for adaptation, not greenhouse gas abatement. If so, Schelling's claim is beside the point. But still, there is a fair question whether current Americans should be forced to make significant transfers because of the decisions of those who preceded them. An approach that emphasized corrective justice would attempt to be more finely tuned, focusing on particular actors, rather than Americans as a class.

Return to the Bosnia-Serbia conflict. Many people who are currently Serbian citizens had no role in planning the genocide and did not benefit from it; some of them opposed the nationalistic policies of Serbia at the time, at great personal risk. Many Serbians today were children or not born during the genocide, and others immigrated after the genocide (some as refugees escaping atrocities in Bosnia). And some people living in Serbia are victims of the genocide, or relatives of the victims, or victims of retaliation by Bosnians. Yet by holding Serbia liable for the genocide, one forces all these people to pay higher taxes. This violates moral judgments against collective responsibility.[33]

The most natural and best response to this point, signaled previously, is that all or most Americans today benefit from the greenhouse gas–emitting activities of Americans living in the past, and therefore it would not be wrong to require Americans today to pay for losses or adaptation (or mitigation). This argument is familiar from debates about slave reparations, where it is argued that (mostly white) Americans today have benefited from the toil of slaves 150 years ago.[34] To the extent that members of current generations have gained from past wrongdoing, it may well make sense to ask them to make compensation to those harmed as a result.

But this argument also raises questions, conceptual and empirical. Many Americans today are, of course, immigrants or children of immigrants, and so not the descendants of greenhouse gas–emitting Americans of the past. It is possible that such people nonetheless benefit from past emissions, but perhaps they have received little or nothing from them. Further, not all Americans inherit the wealth of their ancestors, and even those who do would not necessarily have inherited less if their ancestors' generations had not engaged in the greenhouse gas–emitting activities.

From the standpoint of corrective justice, there is another point, and it is admittedly somewhat uncomfortable. As long as the costs are being toted up, the benefits should be as well. No one should doubt that climate change is imposing costs far in excess of benefits, but it is also expected to produce some benefits, both by increasing agricultural productivity and by reducing extremes of cold.[35] And while past generations of Americans have imposed large climate-related costs on the rest of the world, they have also conferred massive benefits that have not been fully internalized. American industrial activity has produced products that are consumed in numerous foreign countries, for example, and has driven technological advances that have benefited people all over the world. What would the world, or Pakistan, look like if the United States had produced 10 percent of its level of greenhouse gas emissions, or 20 percent, or 40 percent? A proper accounting would seem to be necessary, and it presents formidable empirical and conceptual problems. Should the benefits be subtracted from the costs?

But-for causation arguments used in legal analysis present serious problems when applied historically. We can meaningfully ask whether an accident would have occurred if the driver had operated the vehicle more slowly, but to say the least, it is challenging to answer the question how the rest of the world would look if the Americans of prior generations had not engaged in the activities that produced greenhouse gases. In this hypothetical world of limited industrialization in the United States, Pakistan would be a different country, and the rest of the world would be unrecognizably different as well.

## CAUSATION

Corrective justice requires that the wrongdoing cause the harm. In ordinary person-to-person encounters, this requirement is straightforward. But in the context of climate change, causation poses challenges.

To see why, consider a village in India that is wiped out by a monsoon. One might make a plausible argument that the flooding was more likely than it would otherwise have been as a result of rising sea levels caused by climate change. But it would not be so easy to show that greenhouse gas emissions in the United States caused the flooding, or even contributed to it.[36] If the flooding was in a probabilistic sense the result of greenhouse gas activities around the world, it was also the result of complex natural phenomena that are not entirely understood. And to the extent that the United States was involved, much of the contribution was due to people who died many years ago; in all likelihood, little (or none?) was due to people who engage in activities that produce greenhouse gas emissions today.

Causation problems are not fatal to corrective justice claims, but they weaken them. In tort law, occasionally courts are willing to assign liability according to market share when multiple firms contribute to a harm—for example, pollution, or dangerous products whose provenance cannot be traced.[37] In this light, perhaps policymakers should operate in probabilistic terms and build compensation requirements on the best available understanding of the damage attributable to greenhouse gas emissions by various nations. But statistical relations are not the same as causation, and for some harms, they become too weak to support a claim about corrective justice.

## CULPABILITY

Philosophers disagree about whether corrective justice requires culpability.[38] Frequently intentional, reckless, or negligent action is thought to be required for a corrective justice claim. While some people do support strict liability on corrective justice grounds, a degree of culpability is required to make the analysis tractable: because multiple persons and actions (including those of the victim) are necessary for harm to have occurred, identification of the person who has "caused" the harm requires some kind of assignment of blame. At a minimum, the case for a remedy is stronger when a person acts culpably than innocently, and so it is worthwhile to inquire whether the United States or Americans can be blamed for contributing to climate change. Indeed, the notion that Americans have acted in a blameworthy fashion by contributing excessively to climate change is an important theme in popular debates. Something similar can be explored, of course, for people in other wealthy countries—Canadians, Germans, Italians, Norwegians, and so forth. The underlying issues are not simple.

The weakest standard of culpability is negligence: if one negligently injures someone, one owes the injured party a remedy. Economists define *negligence* as the failure to take cost-justified precautions.[39] Lawyers tend to appeal to community standards.[40] Today, the scientific consensus holds that the climate is warming and that this warming trend is due to human activity. But this consensus took a long time to form. In the modern era, the earliest work on global warming occurred in the 1970s, and it was hardly uncontroversial. At a minimum, greenhouse gas–emitting activities did not become negligent, under existing legal standards, until a scientific consensus formed and it became widely known among the public—a relatively recent occurrence.[41]

What about the U.S. government? Perhaps one could argue that U.S. climate change policy has been culpably negligent. The argument would be that by failing to take precautions that would have cost the United States a lot but benefited the rest of the world (and possibly also the United States) much more, the U.S. government engaged in culpable behavior. That is a complicated matter, and if the U.S. government has been negligent, it was probably not negligent in (say) 1960 or 1970. Further, it is one thing to say that the U.S. government behaved negligently, and quite another to say that the American people should be held responsible. The government itself does not have its own money to pay the remedy; it can only tax Americans. So the question is whether individual Americans behaved culpably for tolerating a government that failed to take actions that would have produced total benefits in excess of total costs.

Many people are inclined to blame the public for the failures of their political system, but there are reasons for caution. A great deal depends on the details. The last example of such a policy was the war guilt clause of the Versailles Treaty, which held Germany formally responsible for World War I and required Germany to pay massive reparations to France and other countries. Perhaps the clause was justified, but Germans resented it, and conventional wisdom holds that their resentment fed the rise of Nazism.

After World War II, the strategy shifted; rather than holding "Germany" responsible for World War II, the allies sought to hold the individuals responsible for German policy responsible—during trials held at Nuremberg and elsewhere, where defendants were given a chance to defend themselves. The shift from collective to individual responsibility was a major legacy of World War II, reflected today in the proliferation of international criminal tribunals that try individuals, not nations.

So it is one thing to blame individuals Americans for excessive greenhouse gas–emitting activity; it is another thing to blame Americans for the failure of their government to adopt strict greenhouse gas policy. Outside of extreme cases, it is fair to question whether voting for politicians who adopt bad policies, or failing to vote for politicians who adopt good policies, is sufficient to justify a claim about corrective justice.

*In short:* Corrective justice intuitions turn out to be an imperfect fit with the climate change problem. I have raised a number of concerns; you might even think of them as objections. In my view, however, the corrective justice argument retains its force, at least if it is presented as the basis of a kind of rough justice in an imperfect world. Wealthy nations (and some not-so-wealthy nations) have imposed serious risks on other nations, many of them poor, and those risks have produced, and will produce, serious and potentially catastrophic harm. Compensation is due. Rough justice is still justice.

## PER CAPITA EMISSIONS

Turn now to a pressing issue. Some developing nations have urged that with respect to emissions reductions, any analysis of national obligations ought to focus on a nation's per capita emissions, not its aggregate emissions.[42] This argument might even be connected with a general right to development, on the theory that a worldwide carbon tax (for example) would forbid poor nations from achieving the levels of development already attained by wealthy nations.

With respect to China, the factual predicate for this argument is that China's population is the largest on the planet, and notwithstanding its explosive emissions growth, its per capita emissions remain well below those of many nations. But the argument for emphasizing per capita emissions might be made by many nations. Consider table 2.7.

Table 2.7
Tons of $CO_2$ equivalents emitted per capita in 2022

| | |
|---|---|
| United States | 19.0 |
| Canada | 18.1 |
| Russia | 14.0 |
| China | 10.1 |
| Japan | 9.4 |
| European Union | 7.6 |
| India | 2.5 |
| Nigeria | 1.9 |

*Source:* Laura Paddison and Annette Choi, *As Climate Chaos Accelerates, Which Countries Are Polluting the Most?*, CNN (Dec. 1, 2023), https://www.cnn.com/interactive/2023/12/us/countries-climate-change-emissions-cop28/.

Poor or not-rich nations might emphasize that their low per capita emissions rate is not only far below that of the United States, but also below those of such nations as Japan, Germany, and the United Kingdom. They might insist that that fact should be taken into account in deciding on appropriate policy. To clarify the claim, assume that the world consists of only two nations, one with two billion people and one with one million people. Suppose that the two nations have the same aggregate emissions rate. Would it make sense to say that for purposes of a system of cap-and-trade, the two should be allocated the same level of emissions rights? Obviously it would not. Many nations therefore argue that all citizens should have a right to the same level of economic opportunity, which means that emissions rights should be allocated on a per capita basis—that nationally determined contributions under the Paris Agreement, or any relevant commitments, should pay careful attention to per capita emissions.[43]

## COMMON BUT DIFFERENTIATED RESPONSIBILITIES

The argument for taking account of per capita emissions is connected with the United Nations Framework Convention on Climate Change's old principle of *common but differentiated responsibilities*.[44] On the surface, this principle means that a nation's obligations on climate issues are to be determined by two factors: its responsibility for climate change and its capacity to cut emissions.[45] Beneath the surface, the principle means that developed nations have to spend a great deal to reduce their emissions, while developing nations do not.

Invoking this principle, many people have called on wealthy countries to take the lead in cutting their emissions and have argued that poorer countries have reduced responsibilities, deserve help for emissions reductions, and are bound to take account of their emissions only as they continue to ensure that their economies grow.[46] Poor nations might contend, and have sometimes contended, that raising the standard of living for their citizens is their first priority.[47] With this point in mind, a poor nation might claim that any actions it takes in regard to climate change will be "within its capability based on its actual situation."[48]

As we have seen, it might also claim, quite plausibly, that wealthy countries have an obligation to assist the developing world with the challenges of climate change, both through technology transfer to allow sustainable development and through financial assistance to adapt to the effects of climate change. As we have also seen, this moral obligation arises because wealthy nations bear the greatest share of responsibility for climate change. Because such nations appropriated more than their share of climate resources in the past, they should now use their wealth to help poor countries develop in or cope with a world where climate resources have become limited.

## ROUGH JUSTICE

If some people are rich and others are desperately poor, those who are rich should help those who are poor. That point supports moral cosmopolitanism of the sort defended in chapter 1. It also supports foreign aid programs, including foreign aid programs designed to help with adaptation to climate-related risks.

It is clear that if one person harms another, compensation is due. This is the argument from corrective justice. Wealthy nations have produced emissions that are endangering poor nations (and less-than-rich nations). It is plausible to think that that they owe them compensation for that harm.

If we put together arguments from distributive justice with arguments from corrective justice, we have a strong basis for asking wealthy countries to help poor countries to deal with climate-related risks. I have emphasized that because countries are not individuals, claims about distributive justice are far from straightforward in this context. I have also emphasized that claims of corrective justice run into legitimate concerns, because we do not have a situation in which clearly identifiable wrongdoers can restore clearly identifiable victims to some clearly identifiable status quo ante. But the fundamental point remains. Wealthy countries are morally obliged to help poor countries manage climate-related risks. Substantial sums are in order to assist with emissions reductions, to promote adaptation, and to provide compensation for harms.

# 3 FUTURE GENERATIONS

Analysis of the ethical issues associated with climate change must come to terms with a simple fact: the benefits of emissions reductions will be enjoyed in the future rather than the present. (Adaptation, by contrast, can help people today.) If the nations of the world dramatically cut emissions immediately, the beneficiaries of that action would be people living some time from now. By contrast, the costs of emissions reductions will be paid mostly by current generations. You might think of emissions reductions as a kind of gift from the present to the future, with the important qualification that if you stop harming people, you aren't exactly giving them a gift.

How should policymakers and analysts deal with future benefits and present costs? The standard answer is that future effects should be "discounted." A dollar today is worth more than a dollar in a year. Should the future benefits of reduced climate change also be discounted?

Discounting the costs and benefits of taking action to prevent climate change may initially seem like a technical, mathematical issue, but it turns out to be one of the central ethical issues in evaluating climate change. Seemingly small changes in the discount rate can lead to very large changes in estimates of the costs of climate change and the benefits of abatement. With a very high discount rate, the argument for immediate, aggressive action to reduce climate change seems weak. With a near-zero discount rate, that argument seems overwhelmingly strong.

Defenders of discounting argue that it is necessary to ensure consistent comparisons of resources spent in different time periods. Critics of discounting begin by insisting on a principle of intergenerational neutrality (endorsed in the introduction). In their view, people in the current generation should not be treated as more valuable than people in the following generations. They object that discounting might ensure that future generations will receive less attention, and perhaps far less attention, than those now living. As we shall see, the issue of discounting turns out to be closely analogous to that raised by the distribution of costs and benefits across countries, except that with discounting, the distribution is across time instead of across space.

## WHY DISCOUNT?

The basic problem addressed by discounting is that the costs and benefits of spending resources to reduce the effects of climate change come at different times. (We are speaking here mostly of mitigation, not adaptation or resilience, though discounting can be relevant to the latter as well.) To prevent significant harms, we have to begin spending sizable resources now and in the near future. The benefits of these expenditures, however, will be enjoyed over the next several hundred years in the form of reduced effects from climate change. Different people will pay the costs and receive the benefits, almost exactly as in the cross-country case.

For short-term projects where the costs and benefits occur in different but relatively proximate periods, it is standard to discount the costs and benefits to a single period. Suppose that you have $100 and are presented with a $100 investment that will produce $110 in two years. You can choose to spend the $100 today or make the investment and spend $110 in two years. You should not simply choose $110 on the grounds that $110 is larger than $100; that would be ridiculous. Instead, you should use discounting to compare the two choices.

To see why, suppose that you have an alternative choice: putting the money in a bank account, which will pay you interest at 6 percent. If you put the $100 in the bank instead of making the investment, you would have $112.12 after two years. Therefore, you should not make the investment. Doing so would leave you worse off than the alternative. The equivalent procedure for making this comparison is to discount the investment returns by the interest rate. If the discounted flow is more than $100, the investment makes sense, and if not, it does not. To discount, simply divide each cash flow by $1/(1+r)^n$, where *r* is the interest rate and *n* is the number of years. Here, the present value of $110 when the interest rate is 6 percent is about $98 ($\$110/1.06^2 = \$97.90$), which is less than $100.

Discounting can be justified on several grounds. In the 2003 incarnation, OMB Circular A-4, a kind of bible for regulatory analysis in the United States, offered three:[1]

1. Resources that are invested will normally earn a positive return, so current consumption is more expensive than future consumption, since you are giving up that expected return on investment when you consume today.
2. Postponed benefits also have a cost because people generally prefer present to future consumption. They are said to have positive time preference.
3. If consumption continues to increase over time, as it has for most of U.S. history, an increment of consumption will be less valuable in the future than it would be today, because the principle of diminishing marginal utility implies that as total consumption increases, the value of a marginal unit of consumption tends to decline.

I am going to have something to say about each of these points. For climate change, the most important point here is 1, which means that discounting is best seen as a way of taking opportunity costs into account. The opportunity cost of making the investment is the alternative return available by investing the money, as by putting the money in the bank. With the numbers given earlier, the opportunity cost was higher than the return on the investment. If the bank paid interest at 4 percent, the opportunity cost would be lower—and, in this case, the investment would make sense.

A key to this analysis is that you have alternative means of shifting resources across time, such as a bank account. Given the bank account, you could choose $100 today or $112.12 in two years, but in no event should you choose $110 in two years. If you were Robinson Crusoe and had no other choice of investments—no way of shifting resources across time periods by borrowing or by saving—the opportunity cost of the investment would be zero and you would simply have to decide whether you preferred $110 in two years to $100 today.

You might still prefer $100 today because you have to eat, but you would no longer look to the opportunity cost to decide.

## THE ARC OF TIME

For climate change, a key question is whether this same logic can be applied over very long time periods (perhaps two hundred years or more) and when the costs and benefits are spread across generations rather than a single individual. Discounting no longer is simply about an individual spreading consumption over a lifetime. Instead, it is about comparing the welfare of different individuals. Does the same logic apply?

It is clear that discounting can have massive effects on policy choices that have long time horizons. For example, suppose that as a result of climate change, we are facing the loss of $1 trillion dollars in one hundred years. If the discount rate is 7 percent, we would be willing to spend only a little over $1 billion—one one-thousandth of the damages—to prevent that harm! If the time horizon were two hundred years—a time well within climate change policy considerations—at a 7 percent discount rate, we would be willing to spend only about $1.3 million, 0.00013 percent of the future cost, to prevent it.

With these numbers, it is easy to construct examples where it is not desirable to spend a small amount of money today to save some valuable asset or large number of people in the distant future. On ethical grounds, many people are skeptical about the conclusions seemingly required by (significantly) discounting the future benefits of reductions in climate change. If people in the distant future would suffer catastrophic losses, and if many of them would die, should we not be willing to spend a good amount to protect them? How can massive losses two hundred years from now be worth very little today?

To see the stakes, consider the fact that a number of years ago, almost the *entire* difference between the influential recommendations of the highly distinguished economists Nicholas Stern and William Nordhaus was driven by disagreement about the appropriate discount rate. Stern discounted future costs and benefits at 1.4 percent, while Nordhaus discounted them at 5.5 percent.[2] A 1.4 percent discount rate would value a cost in one hundred years almost fifty-three times more than would a 5.5 percent discount rate.

If the harm occurs in two hundred years, the Stern approach would value it almost 2,800 times as much as the Nordhaus approach.

To be sure, climate change is causing plenty of harms today. But if the largest costs of climate change occur in the future, then the discounting assumption would naturally lead Stern to see climate change as a far larger problem than Nordhaus does. To confirm that this was the primary difference in the two approaches, Nordhaus ran his computer model using Stern's discount rate. The results with this change echoed Stern's recommendations. To an approximation, discount rates were all that separated the authors![3]

Analysts have long recognized that discount rates are among the central parameters in evaluating the effects of climate change and that the decision about the appropriate response depends on resolving the debate. The International Panel on Climate Change (IPCC), for example, once estimated that discount rates are the second most important factor in evaluating the effects of climate change. The IPCC rated the effect of *climate sensitivity* (there defined as the average global temperature change due to a doubling of the concentration of $CO_2$ in the atmosphere) as the most important factor. If this factor were scaled at 100, discounting would have a value of 66, while estimates of the valuation of the economic impact from a 2.5 degree increase in temperature is valued at 32.[4]

The debate on discounting has a long history. In 1928, Frank Ramsey famously argued, "One point should perhaps be emphasised more particularly; it is assumed that we do not discount later enjoyments in comparison with earlier ones, a practice which is ethically indefensible and arises merely from the weakness of the imagination."[5] That is effectively a claim that if we discount the future, we are acting both unethically and foolishly. Why is a good week now worth more than a good week ten years from now, or one hundred years from now? In a similar vein, Roy Harrod argued that discounting is "a polite expression for rapacity and the conquest of reason by passion."[6] Why is it reasonable to care more about current people than future people?

By contrast, Tjalling Koopmans argued that a failure to discount effectively means that the current generation must starve itself to benefit the future. Suppose that ethically, we must act to maximize total welfare and that a dollar invested grows at a positive rate for the indefinite future. If we do not discount, the future gains from the investment dollar will also be worth infinitely more than the present loss. Therefore, he argued, not discounting means that we must save every dollar, an indefensible conclusion.[7] Countless other authors have studied the issue over the years.[8]

## POSITIVISTS AND ETHICISTS

Broadly speaking, we can identify two major positions with respect to the proper discount rate. *Positivists* attempt to observe what the market-determined discount rate actually is. *Ethicists* attempt to reason from first principles about what the discount rate should be. If the market rate does not coincide with what the ethicists think it should be, the two positions will conflict. In the end, of course, the positivists' approach is worth nothing unless it can be defended on ethical grounds.

As we shall see, Ramsey pointed in the right direction. The ethicists are correct to embrace intergenerational neutrality: the welfare of a person born in 2050 does not matter less than the welfare of a person born in 2010. But the ethicists are wrong to suggest that a commitment to intergenerational neutrality makes it unacceptable to discount money. The positivists are right to insist that money should be discounted. But they are wrong to the extent that they suggest that the welfare of future people, or of people in the future, should be discounted.

As we shall see, these broad propositions do not capture some intricacies. But the conclusion is plain: it is appropriate to use some kind of discount rate, capturing opportunity cost, to evaluate the benefits of reducing greenhouse gas emissions. At the present time, a discount rate between 1.5 percent and 2 percent seems about right, given market rates of return, though specification of the appropriate number, which depends on empirical matters, is not my goal here.

## POSITIVISTS

The positivists approach the issue as a simple problem of opportunity costs, even for the long term.[9] Suppose that we were going to invest $100 billion to reduce carbon emissions, producing a benefit in one hundred years of $400 billion. This represents a rate of return of 1.4 percent, the discount rate used in the Stern review. The positivists reason that if the market rate of return over that time period is 5.5 percent (the Nordhaus rate), the same $100 billion could be invested to produce over $21 *trillion* in one hundred years, almost fifty-three times as much.

Equivalently, we could give the future $400 billion dollars by investing about $2 billion at the market rate, keeping the remaining $98 million to spend on riotous living now. It does not make any sense, argue the positivists, to invest the $100 billion to reduce the effects of climate change under this hypothetical set of facts. To do so would be to throw away vast resources: either $98 billion today or more than $20 trillion in one hundred years.

The conclusion, following this logic, is that it is not sensible to invest in any project unless it has a return at least equal to the return available elsewhere. The problem is exactly parallel to that of the individual who compared the return on the investment to the return available elsewhere (the bank). The long time period does not change the method of analysis; it only makes the issue more important. We should, therefore, discount projects at the otherwise available return—the market rate of return. Only projects that pass discounted cost-benefit analysis should be undertaken. Any other choice throws away resources, which is not at all good for future generations.

In the example given previously, instead of investing the $100 billion in the project that had a 1.4 percent return, we could invest, say, $5 billion in the market and give the future generation about $1 trillion. Everyone would be better off: the current generation would have $95 billion more than otherwise, and the future would have $600 billion more than otherwise.

Note that the positivists are discounting money; they need not argue that they are treating lives at different times equally. They too can accept a principle of intergenerational neutrality. For example, suppose that a statistical life today has a value of $12 million. This is the amount we would be willing to spend to save a life immediately. The positivists argue that if a life in two hundred years is worth the same amount, $12 million (in constant dollars), then we should be willing to put aside only the present value of that amount to save that life. They are not discounting the lives—both are worth $12 million—but they discount the dollars because a dollar put aside today grows with the discount rate.

In this sense, positivists can firmly respect the principle of intergenerational neutrality and can claim that their approach is defensible on ethical grounds. The central argument is that future generations are helped, not hurt, by discounting, because it ensures that resources will be invested for their benefit.

## COMPLICATIONS

This simple analysis runs into three complications.

### UNCERTAINTY

A seemingly technical issue, very much worth attention, involves the effect of uncertainty on discount rates. Suppose that we are considering a project that produces a return of $1 million in fifty years. In addition, suppose that the discount rate is uncertain and can take one of two values: 10 percent or 2 percent, with equal probability. What is the expected discount rate we should use in evaluating this return?

It turns out not to be the simple average of 10 percent and 2 percent (i.e., 6 percent). Instead, the number is far lower—in this case, around 3.4 percent. The reason is that in order to determine the expected discount rate, we need to take the discounted value of the million dollars in each of the two circumstances and average these numbers. If the discount rate is 10 percent, the present value is about $8,500. If the discount rate is only 2 percent, the present value is $372,000. The average value is $190,000. This average, $190,000, is the number we should use for our estimate of the present value of the project. The implied discount rate (i.e., the discount rate that gives a present value of one million as $190,000) is 3.4 percent. As uncertainty increases and as the length of time increases, the effect is magnified.

This point has important implications for the problem of climate change, which is the paradigmatic case of a long-term problem with uncertain effects.[10] This means that the discount rate used by the positivists should be *near the very lowest expected rate of return over the long run.* If an agency uses a high rate, or averages the high and the low, it would be making a serious mistake (and it should be legally vulnerable on the ground that it has acted arbitrarily).

We should discount at the low rates because the bad states of the world—where growth is very low—will dominate the averaging process. Indeed, given that climate change itself might, if bad, lower the rate of return on investments dramatically, the discount rate recommended by the positivists might turn out to be very low. This is true even if the most catastrophic consequences turn out to be unlikely, because the averaging effect just illustrated means that bad outcomes, even if unlikely, dominate the analysis.

### THE RICHER FUTURE

Positivists also have to be sure that they attach the correct values to items in the future. If future people are richer than people alive today, they may value the environment more than people do today; it is a well-known fact that people value the environment more as they get richer. Moreover, if climate change damages the environment (as expected), the benefits it provides will be scarcer, and its relative value increased. Estimating the correct values of the environment in the far distant future will not be easy. But unless we are careful to take into account these sorts of considerations, we are at risk of using the wrong valuations and calculating the costs and benefits of climate change abatement incorrectly.

### PRIVATE AND SOCIAL RATES OF RETURN

A number of technical issues must be addressed in computing the opportunity cost. One important issue turns on how to adjust for taxes and similar items, which cause a divergence between the rates of return that investors see (the after-tax rates of return) and the rate of return that benefits society (the full, pretax rate of return, because the investor gets the after-tax amount and the government gets the taxes). Another problem is that there are many market rates of interest. Treasuries pay a different rate from corporates; short-term bonds pay a different rate from long-term bonds; stocks and bonds have different returns. While important, these issues are largely technical and need not detain us here.

## ETHICISTS

The ethicists argue that the only way to determine the correct discount rate is to go back to first principles of ethical reasoning. They insist that cost-benefit analysis with discounting can result in clearly unethical choices. Climate change provides the most vivid example because it exposes the future to a serious risk of catastrophic harm. The discount rates potentially required by the positivists' approach might make us unwilling to spend a relatively small amount today to prevent these serious harms in the future.

If we respect a principle of intergenerational neutrality and believe that we have an ethical obligation to take the interests of members of future generations seriously, this is unethical. Discounting cannot justify refusing to spend small amounts to prevent causing the risk of terrible harm to others.

To be more concrete, suppose that sea level rise will destroy Florida in two hundred years. Suppose also that these effects will be very difficult to counteract: reducing emissions may be extraordinarily expensive, and so is adapting to the resulting harm. Cost-benefit analysis using the market return as a discount rate might recommend that we do very little. This, however, says nothing about whether we are behaving justly toward our descendants. We may, under a variety of ethical theories, owe them prevention of or compensation for this harm.

The philosopher John Rawls has argued that the generation in which one happens to find oneself is irrelevant from the moral point of view; the current generation violates its moral obligations if it enriches itself while subjecting future generations to catastrophic harm.[11] Under a welfarist approach, the question is whether the actions of the current generation increase or decrease overall welfare; it is easily imaginable that a failure to take certain actions to prevent climate change could cause a far greater welfare loss, to all generations taken as a whole, than would those actions themselves. The positivists' theory of discounting has nothing to say about the obligations of the current generations. In short, choosing projects solely through cost-benefit analysis with discounting can result in serious injustice and may violate our ethical obligations to the future.

By examining the details of discounting, the ethicists show why it produces what seem to be obviously unethical results. They offer three central reasons.

### PRIVATE VERSUS SOCIAL RATES OF RETURN

The ethicists argue that the rate of return on an investment seen by individuals—the so-called private rate of return—is not the same as the benefit society gets from an investment—the social rate of return. There could be many reasons for this, but unless markets are perfectly competitive, a condition that is unlikely to hold, the two will not be equal. Observed interest rates reflect only private rates of return and, therefore, they are not a good guide to whether the total, social benefits from an investment are worth the costs.

To illustrate using the discount rates discussed previously, individuals may demand a 5.5 percent rate of return on investments, but they see only their private benefits. The social benefits of an investment may be much greater. It may, for example, be the case that if individuals get a 1.4 percent benefit, the additional benefits to society from an investment make it worthwhile. Looking only to the market rate, 5.5 percent, would mean that we reject projects that overall are worthwhile. Absent very stringent conditions, such as the economists' imaginary perfect market, the private and social benefits will not be equal, so we cannot look to the private returns available in the market as a guide for social policy.

### INDIVIDUALS NEGLECTING POSTERITY

Individuals determine today's interest rate by deciding how much to save for the future. In making this decision, individuals are considering their *lifetime consumption*—how much to consume today compared to their retirement—and possibly the consumption of their children or grandchildren. Climate change, however, is a problem that will last many hundreds of years, spanning multiple generations. Individuals, the ethicists claim, are simply not thinking about the distant future when making savings decisions. This means that observed interest rates are not a good guide for decisions over very long time periods. Individuals, in setting the market rate of return, are simply not considering the relevant question.

### CHANGING RATES OF RETURN

Finally, climate change abatement will involve large adjustments to the economy. If we make these adjustments, rates of return are likely to change. That is, the rate of return to investments is endogenous to the problem, not something that is external to the problem. It is a variable we choose rather than a variable we observe. To make this choice, we have to be explicit about the reasons. We have to go back to ethics.

## A THOUGHT EXPERIMENT

Starting from scratch to try to determine the appropriate discount rate, the ethicists imagine society conducting a thought experiment. Here the relevant claims become somewhat technical, and to understand them, it is necessary to introduce a little math. Imagine that we are considering investing an additional dollar today that will produce returns in the future. How much must that return in the future be to compensate for the reduction of consumption today? The difference—how much the future must get if the present loses a dollar—is based on our ethical judgments about how much each generation deserves. The *social discount rate* is this net increase in consumption. So if we must give up one unit today and society would demand $1 + \rho$ in the future to offset this loss, $\rho$, the Greek letter rho, is the discount rate.

The ethicists assume that society determines the answer to this question by maximizing the sum of the welfare of each generation, possibly discounting for time in this process. That is, in symbols, society wants to find the following:

$$\text{Max } [W(C_0) + W(C_1) / (1 + \delta) + W(C_2) / (1 + \delta)^2 \ldots]$$

Where $\delta$ is the rate at which we discount the welfare of future generations, if at all, and $W(C_i)$ is a measure of the welfare of a given generation, $C_i$ being the consumption of that generation.

The ethicists then make one additional special assumption. They assume that the welfare of each generation is determined by a very particular form: as consumption goes up, the marginal benefit of additional consumption (in percentage terms) goes down at a constant rate, $\eta$, the Greek letter eta. In particular, the welfare of a given generation, W, is set equal to $C^{(1 - \eta)}/(1 - \eta)$. This is largely chosen for convenience rather than based on any empirical or ethical support; this functional form happens to be easy to work with mathematically.[12]

With these two very specialized assumptions—that society uses a particular function to choose how to distribute wealth across generations and that it weighs each generation using a specialized functional form which happens to be simple to use for calculations—and some simple math, the ethicists derive the following as an expression for the social discount rate:

$$\rho = \eta\dot{c} + \delta$$

The explanation of this formula is straightforward. As discussed, $\rho$ is the social discount rate—that rate we should use for evaluating projects such as climate change abatement. The term $\dot{c}$ is the rate of growth of the economy. It tells us what the consumption in each period, the C elements in the formula, will be. Eta reflects how much we care about inequality. That is, suppose that growth rates are high, so that the future is very much richer than today. Eta tells us that an increment to their consumption matters less because they are richer. The higher eta is, the more we demand the future get for us to give up a unit of consumption today. We can think of it as an inequality parameter.

The second term is $\delta$, the rate at which we discount future generations. The ethicists call this the *pure rate of time preference*. It reflects a lower evaluation of future generations simply because they are distant from us.

Most of the discussion has been about $\delta$. The ethicists almost uniformly take the position that it is unethical to allow $\delta$ to be positive because this would mean giving less weight to a future person simply because he lives in the future. If all people count equally, $\delta$ must be zero. (That sounds right!)

Eta is also important. It reflects views about inequality, both across generations and within any given generation. Higher eta values are more egalitarian. In the climate change context, the more egalitarian we are, the higher the discount rate and the less we should be willing to invest in abatement. If we use the same values for redistribution with the current generation as across time, the more we want to redistribute now (i.e., we are more egalitarian), the higher the discount rate should be (we should less want to engage in climate change abatement because it redistributes toward the richer future). This seeming paradox, that strong egalitarians should care less about the future, arises because the future is expected (we hope) to be richer than today.

The terminology in the debate, at this point, becomes a little confusing. Setting $\delta$ equal to zero is often described as *not discounting*. In a sense this is correct: future generations are not discounted merely because they are in the future. The pure rate of time preference is zero. The social rate of discount, $\rho$, however, will be positive, but this, the ethicists say, is not because we are discounting. It is because we have distributive preferences: to the extent that if the future is richer, we should less want to increase their consumption. If $\rho$ is positive, however, there will be a mathematical procedure used in evaluating climate change that is identical to discounting. The procedure, however, is about adjusting for distributive preferences, not time. It looks like discounting only by coincidence.

Once we have had our ethical debates about the parameters (and made good technical estimates of the growth rate, $\dot{c}$), we can determine the social discount rate. Stern in his initial report set $\delta = 0.1$, $\eta = 1$, and estimated the growth rate as 1.3 percent. Therefore, he used a social discount rate of 1.4 percent. Note that this was well below the rate of return available for investments in the economy. He later modified his views, adjusting $\eta$ up to 2, so that the discount rate would now be 2.7 percent, closer to but still below the rate of return on other investments.[13] In the same period, Nordhaus would have set $\delta = 1.5$, $\eta = 2$, and a growth rate of 2 percent, to derive a discount rate equal to 5.5 percent. The difference in conclusions from using these different discount rates is, as noted, dramatic, shifting the policy recommendation from among the most conservative to the most aggressive.

## THE ETHICS OF DISCOUNTING

Now consider two propositions:

1. The ethicists are correct to insist we must respect the principle of intergenerational neutrality.
2. The positivists are correct that choosing any project that has a lower rate of return than the market rate of return throws away resources, and in a way that threatens to harm future generations. It follows that the ethicists' insistence on intergenerational neutrality does not justify rejecting discounting at the market rate of return (properly taking into account the various qualifications noted earlier, including uncertainty).

To illustrate how these two propositions can be separated, imagine that the current generation is leaving a given legacy for the future, a legacy that for now we imagine to be the ethically justified amount. In monetary equivalents, call it one hundred dollars. Suppose also that a new project is being considered that costs the current generation ten dollars and produces twenty dollars for the future. If we engage in this project, we can reduce our legacy elsewhere, still leaving one hundred dollars for the future, maintaining the requisite neutrality. The only question in this case is whether spending the ten dollars on this project produces a better return than spending the ten dollars elsewhere. The correct procedure for deciding whether to engage in this project is to measure the opportunity costs, which, as showed earlier, is equivalent to discounting.[14]

Now suppose that we find out that our legacy to the future is inadequate because of newly discovered environmental harms from our actions.

Suppose, for example, we discover that it is only seventy dollars instead of one hundred dollars. We must now reevaluate whether we are leaving enough. If we believe that, given this information, the correct amount to leave to the future is ninety-five dollars, then we must increase our legacy.[15] We should do so in a way that costs us the least, which means considering the opportunity costs of alternative projects. If we can find a project that costs us only ten dollars and leaves twenty-five dollars for the future, we should not engage in projects that cost more in return for that amount.[16] The market rate of return measures the returns from currently available projects, so as an initial matter, the market rate is a measure of the opportunity costs of this choice. Once again, therefore, we should use discounting at the market rate to choose projects. Project choice and ethical obligations to the future are, to a large extent, separate.

Seen this way, the ethicists' criticism of the positivists' opportunity cost argument is simply irrelevant. It does not matter whether the current market rates of interest are ethically correct because they still represent the opportunity costs of investment. Recall the numbers used earlier: if we could invest $100 billion to produce $400 billion of benefits in one hundred years when that market rate is 5.5 percent, we could equivalently invest only $2 billion to produce those same benefits.

The ethicists argue that the 1.4 percent rate of return on the $100 billion (to produce $400 billion) is good enough once we consider the ethical arguments. That is, ethical considerations show that society should make investments not only with a 5.5 percent rate of return but also with a 1.4 percent rate of return. But given that there are hundreds or thousands of investment choices, if we are going to make investments at less than the 5.5 percent rate, we should start with those with the highest return. That is, at least initially, the opportunity cost is 5.5 percent. If we make enough investments to exhaust the opportunities at this rate of return, we can begin moving down the scale, but in no case should we jump to investments with very low rates of return as a first option.

This point is central: the ethicists' arguments are that we are leaving an insufficient amount for the future given current policies. On certain assumptions about the effects of current decisions on the future, these arguments are correct. But regardless of whether it is, it says nothing about the particular choice of projects or policies. If we are going to increase the amount we leave for the future, it is incumbent on us not to do so in a way that wastes resources. Therefore, even if the ethicists' argument is entirely correct, we still must carefully consider the opportunity costs of projects and pick those with the highest return.[17]

The positivists, however, also make a mistake. As we have seen, using a market discount rate (properly adjusted for uncertainty) is not a reason for failing to discharge our obligations to the future. The underlying intuition behind the ethicists' argument is that current policies threaten to impoverish the future or to reduce greatly its welfare (as climate change threatens to do). If this is true, discounting is not a reason for allowing this to happen. It is simply a method of choosing projects that fulfill our obligation to prevent this from happening. A recommendation for relatively modest climate change abatement may also need to be accompanied by other projects that ensure the proper intergenerational distribution of welfare. That is, the ethicists may very well be correct that we need to adjust the amounts we are leaving for the future in light of our new understanding of the effects of climate change, while the positivists are correct that in doing so, we must be sure to pick those projects with the highest rates of return. Climate change abatement would be justified if and only it counts as such a project.

## OBJECTIONS

Consider here three objections to the claim that the positions of the ethicists and positivists address separate issues.

### THE LINK BETWEEN THE MARKET RATE OF RETURN AND OVERALL SAVINGS RATE

The ethicists will object that it is not possible to distinguish between the market rate of return and our ethical obligations to the future. The reason is that if we were to save more, market rates would go down. When we finally are saving the right amount, the market rate will be the rate prescribed by their ethical arguments. Therefore, we might as well choose projects with that rate of return.[18] Similarly, as mentioned earlier, the ethicists might argue that the market rate of return is a choice variable, not an exogenous input, when we are dealing with large projects. The separation of the market rate of return and the choice of projects does not make sense if the market rate of return is determined by the choice of projects.

While there is a likely connection between the overall savings rate or very large projects and the market rate of return, the ethicists' conclusion does not follow. If the market rate of return is, say, 5.5 percent and the ethicists argue that at the correct savings rate, the rate of return should be 1.4 percent, then we should not immediately jump to projects with such low rates of return because eventually, if we increase savings enough, market rates of return might get this low. Large adjustments to our legacy to the future are difficult and their success unclear. We should begin by choosing high-return projects, not low-return projects.

Moreover, it is probably wrong to suggest that if we sufficiently increase our savings, interest rates will equal the one determined by the ethicists' intuitions. The basic macroeconomic model used by the ethicists to derive their equilibrium market rate of return is the Ramsey model, which has been supplanted by vastly more sophisticated models.[19] Even the most sophisticated and modern models cannot compute equilibrium interest rates when there are large-scale changes to the economy, such as a vast increase in savings. It does not seem wise to make decisions by relying on an outdated model to argue for committing potentially trillions of dollars to a project on the theory that in the eventual equilibrium predicted by that model, the project choice will seem sensible.

A better decision procedure is to consider, more directly, the nature of our ethical obligations to the future. If the obligation is to leave more to the future than we previously thought (say, because of newly discovered risks from climate change), we need to decide how to do that. The first choice of projects as we begin this adjustment should be those with the highest rate of return. This means discounting at the market rate. As market returns adjust (by going down if the model used by the ethicists is correct), the opportunity cost of new projects goes down.

## FEASIBILITY

While in theory the ethicists' and positivists' positions can be reconciled, the most difficult problems for both are potentially ones of feasibility. Let us start with the feasibility problems of the ethicists' position and then turn to the problems of the positivists' position.

The ethicists derive a discount rate to be used by a social planner independent of the rates demanded by individuals. That is, they start off with the basic premise that social rates of return and private rates of return are different and that the government should choose projects using the social rate of return. The problem with framing the question in this way is that individuals control most of the wealth in society. This means that whatever their preferences about savings, even if wrong, they can offset whatever the government does.

Suppose, for example, that individuals, taken as collective, want to leave one hundred dollars for the future and are doing so now. On the basis of the ethicists' recommendation, the government invests in a new project that leaves forty dollars for the future. Individuals seeing this project can simply reduce their legacy to sixty dollars and keep the total at one hundred dollars, frustrating the government's attempt to correct the market. If individuals can make these adjustments, the question the ethicists start with is simply the wrong question, because the government is not making the choice that is posed. Instead, the government is merely choosing which projects will be included in the total amount left for the future.

This type of behavior is known in economics literature as *Ricardian equivalence*. The extent to which individuals behave this way is much debated and it is unlikely to be fully true.[20] Individuals may, under our numbers, reduce their legacy to only seventy dollars or eighty dollars or even ninety dollars. Nevertheless, there is a basic futility problem with the ethicists' approach. The question they pose, by imagining that the government makes the basic choices about total investment, ignores the basic (and well-founded) constraints on the government.

The positivists also have feasibility problems. The problem is that it may not be possible to transfer resources across hundreds of years to compensate the future victims of climate change. We are, when it comes to climate change, like Robinson Crusoe: the choices are simply about distribution because there is no "bank"; there is no other way of shifting resources across these long periods of time. The point is fundamental.

This argument has been made most convincingly by Robert Lind.[21] He imagines a proposal to transfer resources to the distant future through an investment with a 0 percent rate of return at a time when money or other projects earn a 10 percent return:

> The preferred decision may well be to make that investment and transfer the resources to the future generation even though it earns a zero rate of return. At this point an eager graduate student jumps up, sensing an economic slam dunk,

> and says "That was a really dumb decision. You could have invested that money at 10% and made those people a lot better off." Wrong! We don't know how to set aside investment funds and to commit intervening generations to investing and reinvesting those funds for eventual delivery as consumer goods to the generation 200 years from now.[22]

The extent to which this is correct is an empirical and institutional question rather than a purely ethical question. We cannot rule out the possibility that projects with very low rates of return are the best way of shifting resources across time, but it seems unlikely. If, for example, the market rate of return were really 10 percent and the project at issue had a zero rate of return, it is hard to imagine that there were not other projects that, while perhaps not yielding the full 10 percent, had a higher rate of return than zero.

The claim that we cannot set aside funds for the future is often based on the problem of intervening generations. Suppose that a project, such as climate change abatement, will pay off in the long-distant future—say, in one hundred to two hundred years. If we try to set aside funds for those same future individuals on the theory that the set-aside funds will have a greater future value than funds invested in reducing emissions, they would have to pass through many generations before being received by their target. Any of those intervening generations could prevent the transfer to the future, making it impossible to guarantee that the funds will be used as intended.

Note, however, that the same problem arises with climate change abatement. Even if we spend vast resources reducing carbon emissions, future generations can always revert to burning fossil fuels. It is hard, without much more institutional detail, to understand why various projects would differentially have the problem of intervening generations, which is what is needed for the claim to have force. This possibility can be ruled out, but it seems extremely unlikely.

## INCOMMENSURABLE GOODS: LIVES VERSUS MONEY

A final argument for why we cannot separate our ethical obligations and project choice is that climate change produces particular harms to the future that cannot be offset by other projects that help the future in other ways. This argument may take a variety of forms. One version is that the deaths caused by climate change cannot be offset by simply saving more. Richard Revesz makes such an argument.[23] He contends that the primary reason for discounting monetary benefits does not apply to risks to life and health. Money is discounted because it can be invested. But human lives cannot be invested, and a life lost twenty or two hundred years in the future cannot be "recovered" by investing some sum in the present.

Revesz is right to say that lives cannot be invested, but what is being discounted is money, not lives. Under the standard analysis, any discount rate applies to willingness to pay to reduce statistical risks, which is a monetary measure.[24] The issues raised by the monetary valuation of lives is no different when used for standard cost-benefit analysis within a single time period and when used over differing time periods. Once lives (or, more properly, statistical risks of mortality) are converted to monetary equivalents, all of the arguments discussed earlier concerning discounting apply. If a life today and a life in two hundred years are both "worth" the same amount in terms of money, we need to discount the dollars allocated to the future life, because money put aside for the future grows.

For example, if a life today and a life in two hundred years are both worth $12 million, we should only allocate the present value of $12 million to the future life. Anything more than that would value the future life more than the present life. To be sure, translating lives into money is not easy and raises a host of thorny questions (see chapter 4, but it is not an issue for discounting in particular; objections to the methodologies for valuing lives are orthogonal to the discounting debate).

An alternative version of the incommensurability argument, associated with Derek Parfit, is that even if discounting combined with changes in overall savings rates can get the allocation of resources correct across generations taken as a whole, harms to particular individuals cannot be offset by this procedure.[25] Parfit imagines activities today that increase the risk of genetic deformities in a small number of individuals in the future. Overall changes to the allocation of resources across generations do not compensate those individuals.

In a sense, Parfit is right, but just as with the problem of valuing lives, this argument is not really about discounting. A project that takes place entirely within a single time period may still impose risks on individuals. Before the project is begun, all individuals may be subject to the same risk, and taken as whole, the project may seem sensible. Ex post, however, particular individuals will suffer harm, and those who gain from the project may not be able to compensate those who lose. The arguments surrounding this problem have been debated vigorously. It presents nothing new when the project occurs over more than one time period.

## RESPECTING THE FUTURE

The minimal goal of this chapter has been to illuminate the debate over discounting and intergenerational justice—to show exactly what is dividing the two sides. On the most fundamental question, ethicists are right to insist on a principle of intergenerational neutrality. Welfare should not be discounted. But for purposes of practice, positivists are largely right. Money should be discounted.

It follows that projects, including those involving climate change, should be evaluated by discounting the costs and benefits at the market rate of return, properly adjusted for uncertainty and for the value of the environment. Any other approach risks choosing projects with low rates of return, leaving resources on the table and likely harming the future.

Discounting should be seen only as a method of choosing projects, not as a method of determining our ethical obligations to the future. We should insist on a principle of intergenerational neutrality. If climate change means that what we leave to our descendants is far less than we hoped and thought, then we have a moral obligation to adjust. The proper response is to leave them more, not to choose projects by refusing to discount.

# 4 VALUING LIFE: WHO WINS, WHO LOSES?

We have seen that in deciding how much to do about the risks of climate change, and in specifying the social cost of carbon, the government has to value human lives. If some action will save one hundred lives, or five hundred lives, or one thousand lives, what is its benefit? That might seem to be a horrifying question, and in many ways it is. But there is no way to avoid answering it, which means that we are immediately in challenging ethical terrain.

In policy circles, there is a great deal of reliance on the *value of a statistical life* (VSL), which is a bit like toothpaste. Many people use it, but no one really loves it.

The problem might lie in its (misleading) name. Who loves the idea of putting a monetary value on a life, even if it is merely a statistical one? If we spoke more precisely of the *value of a statistical mortality risk*, the idea might be a little more lovable, or at least a little less unlovable. Still, it is not a lot of fun to think about mortality risks, and if we are turning them into monetary equivalents, well, who could love that? Perhaps especially in the domain of climate change, where we have to think of variations across nations?

For a glimpse of ordinary practice, here are some words from the Office of Management and Budget: "Some describe the monetized value of small changes in fatality risk as the 'value of statistical life' (VSL). This term refers to the measurement of willingness to pay for reductions in only small risks of premature death. It has no application to an identifiable individual or to very large reductions in individual risks. It does not suggest that any individual's life can be expressed in monetary terms."[1] The basic idea here is that public officials have to assign *some* monetary value to mortality risks, even if they do so only implicitly. If they are deciding on one level of stringency and not another, they are necessarily deciding that that level of stringency is worthwhile (and that another level of stringency is not worthwhile). The question is not whether to have a VSL; it is what VSL to have. It is important to underline that point, and (for those who object to the whole enterprise) to shout it from the rooftops.

Within the federal government in the United States, VSL is the workhorse of cost-benefit analysis, certainly for purposes of regulatory policy; it is the principal driver of benefits in multiple domains, whether we are speaking of highway safety, road safety, food safety, cigarettes, pandemics—or climate change. That point in turn raises a question: *In deciding on mortality risks, should agencies use a unitary VSL, or should VSL vary with wealth?* Should poor people be taken to have the same VSL as rich people? Is that immoral?

Within the federal government in the United States, the general practice is clear: VSL is unitary. That is true for risks from climate change as well as for risks of all other kinds. In recent years, VSL was about $12 million—period. If a regulation saves ten rich people from dying, its benefits are $120 million, and the same is true if it saves ten poor people from dying. Offhand, the uniform value seems to be justified on the ground that how much protection you get should not depend on how much money you have.

I will raise questions about that justification in due course, but for now consider a simple fact: for the social cost of carbon, a final report from the EPA *weighs mortality costs in proportion to the per capita income of the nation in which people live.*[2] It follows that a Russian life is worth two Ukrainian lives; that a Canadian life saved is worth over sixteen times as much as a Haitian life; that a Qatari life is worth 118 Burundian lives; that a German life is worth twelve Cambodian lives; and that an Australian life is worth four Indonesian lives. Does that make any sense? Is it morally acceptable? Would it be better if the United States treated every life as if it were worth $12 million?

## A LITTLE GUIDED TOUR

To approach these questions, let us turn to a very brief account of existing practice. To produce monetary amounts for statistical risks, agencies in the

United States rely on two kinds of evidence.[3] The first and most important involves real-world markets, producing evidence of compensation levels for actual risks.[4] If employers impose a workplace risk on their employees, how much are employees paid to assume that risk? How much do employees demand? How much do they get? In the workplace and in the market for consumer goods, additional safety has a price; market evidence is investigated to identify that price. Eliminating a risk of 1 in 100,000 is not worth an infinite amount; it is also not worthless. The advantage of real-world markets is that under certain assumptions, they will reveal people's actual preferences, especially when large populations are aggregated. In part for this reason, real-world markets provide the foundation for actual government practice. (We will get to climate change soon.)

A potentially serious disadvantage of these studies is that the underlying decisions are "noisy." A decision to take a construction job, for example, is based on a whole host of considerations. Workers are most unlikely to think: "I will face an annual mortality risk of 1/100,000 rather than 1/200,000, but it's worth it!" How many people know the risks, and how many people balance them against other factors? Whenever people take a job, live in a city, or purchase products, it is not easy for them to isolate the particular component that is attributable to mortality risks. In fact, that seems like a ridiculously impossible task. Consider crime, flooding, air pollution, extreme heat, and drought (and note that all of the items on that list might be associated with climate change).

For this reason, we might object that real-world data about people's choices, involving a large set of factors, cannot possibly tell us how much people are willing to pay, or to accept, for mortality risks of particular sizes. There are related questions about whether people are sufficiently informed and whether their decisions are fully rational. For now, consider the standard response by those who rely on real-world evidence. Across large populations, we really do find sufficiently steady numbers, justifying the view that people are paid a specified amount to face mortality risks. Maybe they are right; maybe they are wrong. To keep things simple, let's just assume that they are right.

The second kind of evidence comes from contingent valuation studies, which ask people, through surveys, how much they are willing to pay to reduce statistical risks.[5] For example: How much would you be willing to pay to avoid a 1/100,000 risk of getting cancer as a result of arsenic in drinking water? Or: How much would you be willing to pay to avoid a 1/100,000 risk of dying from extreme heat?

The advantage of contingent valuation studies is that they can isolate people's willingness to pay to avoid mortality risks. In this respect, they are far less noisy than real-world data. If we find that the average person is willing to pay $100 to eliminate a mortality risk of 1/100,000, perhaps we can conclude that VSL is $10 million—not in the sense that a life is really "worth" that amount, or even in the sense that a 1/1000 mortality risk is worth $10,000 or a 1/100 mortality risk is worth $100,000, but in the sense that when government is eliminating a 1/100,000 risk faced by a large population, $100 is the right number to assign. (We are bracketing for the moment the question whether poor people should be treated differently from rich people.)

But there are serious disadvantages to contingent valuation studies. The questions are hypothetical and highly unfamiliar, and there are many reasons to wonder whether they provide an accurate measurement of something, rather than a stab in the dark.

Suppose that you are asked how much you are willing to pay to eliminate a mortality risk of 1/10,000. Would you say $0? Would you say $200? Would you want some information first? The question might seem so odd and so confusing, and possibly so unwelcome, that any answer you give will tell us very little. Indeed, researchers do tend to get puzzlingly diverse answers. In my own studies among college students, the average answer is often in the general vicinity of the government's current number (recall that it is $12 million, suggesting a willingness to pay $120 to eliminate a mortality risk of 1/10,000), but the average conceals a lot of individual differences. Some people say that they are willing to pay $0 to eliminate a risk of 1/10,000. Some say that they are willing to pay $200 or more, suggesting a VSL that ranges from $0 to $20 million or higher. Perhaps the diverse answers reflect different risk preferences and different levels of wealth. But perhaps they reflect confusion.

There are other puzzles. How much would you pay to buy safety equipment to eliminate an annual risk of 1/100,000 from driving your new car? In informal studies, I have sometimes obtained a high amount from that question—an average in the general vicinity of $250. How much would you pay to eliminate an annual risk of 1/100,000 from contaminants in drinking water? In informal studies, I have sometimes obtained a low amount from that question—an average in the vicinity of $50. How much would you pay to eliminate an annual risk of 1/100,000 from extreme heat? No one should be surprised if the number turns out to be exceedingly high. The disparities may reflect a belief that context matters, and that all 1/100,000 risks are not the same. Or they may reflect confusion, as when a new car purchase (requiring an expenditure of $20,000 or more) biases people in the direction of providing high numbers for risk reduction.

But let's bracket the complexities here.[6] For government, the relevant risks usually are in the general range of 1/10,000 to 1/100,000. The calculation of VSL is a product of simple arithmetic. Suppose that workers must be paid $1,200, on average, to assume a risk of 1/10,000. If so, the VSL would be said to be $12 million.

## RICH AND POOR

Now let us turn to the question how to deal with differences between rich and poor. You can think of the two as individuals, or you can think of them as communities, or you can think of them as nations.

For orientation, suppose that there are only two kinds of cars: safe cars and safer cars. Suppose also that there are only two kinds of people: poor people and rich people. Safer cars are more expensive than safe cars. Let us suppose that poor people buy safe cars, not safer cars, because they are not willing to pay the (high) price of safer cars. Let us suppose too that rich people buy safer cars, not safe cars, because they are willing to pay the premium. Under imaginable assumptions, we might want to subsidize safer cars so that everyone can buy them (though as we have seen, it might be better to make a lump-sum payment to the poor, so that they can buy what they most want and are not told that they must spend the money on safer cars). That point has implications for climate change, of course.

But maybe subsidies are not feasible, and maybe they are not a terrific idea. (After all, safe cars are safe; they are not as safe as safer cars, but they are safe.) If subsidies are off the table, it is not so easy to see why we might want to require everyone, poor and rich, to buy safer cars. Why should we force poor people to buy cars that they do not want? Shouldn't we acknowledge that if they do not want to buy safer cars, it is probably because they want to use their limited resources for something else? If we care about people's welfare, we ought not to require them to buy something they do not want. If we care about people's autonomy, we ought to respect their wishes.

To be sure, there might be good reason to think that in deciding what to buy, poor people are making mistakes. For example, they might be undervaluing safety. Perhaps it could be said that poor people lack relevant information or suffer from some kind of behavioral bias (such as present bias or limited attention), such that a regulatory mandate would be in their interests. We cannot rule out that possibility. But the claim would have to be demonstrated, not simply asserted, and no absence of information or behavioral bias could support the view that poor people ought to be *required* to spend as much on motor vehicle safety as rich people are willing to spend on motor vehicle safety.

A great deal of empirical work uncovers the unsurprising fact that to reduce statistical risks, poor people are willing to pay less than rich people.[7] Similarly, people in poor countries are willing to pay less than are people in rich countries. We have seen that the American population-wide VSL is about $12 million, while in India, it was about $1 million in 2018.[8]

The empirical findings fit with the simple claim that if you have a lot of money, you are willing to spend more to reduce statistical risks than you would be if you have little money. The findings are also in line with a long-standing claim in utilitarianism, which is that a given unit of money is worth more to a poor person than to a rich person.[9] If the five richest people in the world were given $5,000, the gift would matter little if at all to their welfare; if the five poorest people in the world were given $5,000, the gift would make all the difference in terms of their welfare. By the way, it follows that there is a strong argument in favor of redistribution on simple utilitarian grounds (bracketing incentive effects). Redistribution through taxation might seem the best and most natural response.[10] But if the tax system is unavailable for that purpose, for political or other reasons, perhaps there are other routes.[11]

## THREE CLAIMS

Consider in this light three claims that bear directly on the ethics of climate change:

1. Whether a uniform VSL is in the interest of poor people, or people in poor nations, depends in the first instance on whether we are dealing with subsidies or regulations.

2. In the case of *subsidies*, the use of a VSL that reflects a worldwide average, or even the number for the top quartile, would be very much in the interest of poor people and people in poor nations; in all probability, they would be net gainers. If the social cost of carbon is based on a $12 million VSL for everyone, poor countries would be better off, because they would get bigger benefits (in the form not of cash, but of reduced emissions from the United States).
3. In the case of *regulations*, a uniform VSL is more likely to benefit poor people if they do not pay all or most of the cost; it is more likely to harm them if they pay all or most of the cost. To know the effect of a uniform VSL on poor people in the case of regulations, it is essential to know the *incidence* of the costs (and also the benefits). Who is paying? Who is benefiting?

Let us anchor the discussion with a set of stylized problems, understood to be simplified (but not unrealistic) descriptions of real-life situations.[12]

## PROBLEM 1

The U.S. government is funding various programs to diminish risks, including mortality risks, some of which stem from climate change (extreme heat, drought, wildfire, flooding).[13] Some of the programs are domestic (funded, say, by FEMA). Some are international (funded, say, by USAID). To receive funding, communities that apply for funding must show that the benefits of their programs justify the costs. Some of those communities are poor; some are wealthy. In some poor communities, the average VSL is $1 million. In others, it is $250,000 or even less. In the wealthy communities, the average VSL is $20 million. In its funding decisions, the U.S. government uses a population-wide average for the United States, which is $12 million.

Use of the population-wide average is highly likely to be in the interest of poor communities. The government's policy gives them an in-kind subsidy: They might be willing to pay ten dollars to avoid a mortality risk of 1 in 100,000, but the government acts as if they are willing to pay one hundred dollars. Because the government is footing the bill, people in poor communities in the United States and elsewhere are unlikely to lose. True, they might prefer the money as an unrestricted cash transfer to the boosted

VSL in a risk-reduction program. But they can only gain, and cannot lose, from treating their VSL as if it were $12 million.

The major qualification is that the money for the grant must come from somewhere, and we cannot rule out the possibility that its use here will mean it will be unavailable for some other program or initiative from which poor people would benefit more. Does the program here entail higher taxes on the rich? Spending cuts from programs that help the poor in the United States and elsewhere? Deficit spending? If we know that the funding does not entail some loss for the relevant community, or for poor people in general, the optimistic conclusion holds.

For comparison: Suppose that property values, in poor communities, are low, such that such communities cannot show high monetized benefits from protecting property against climate-related risks in those communities. Suppose that the government believes that the monetized benefits understate the welfare benefits. In a rich community, the average home might be valued at $3 million, and in a poor community, the average home might be valued at $200,000. The *monetary* loss is much lower in the poor community, but the *welfare* loss from the destruction of the average home in the poor community might be equal to or greater than the welfare loss from the destruction of the average home in the rich community.[14] If the government puts some kind of distributional weight on the average home in the poor community, to capture that welfare loss in the context of a subsidy program, it is helping members of that community. As we have seen, using a distributional weight is a good idea on welfare grounds. It is also a good idea on distributive grounds.

## PROBLEM 2

The U.S. government is regulating motor vehicles to diminish safety risks, including mortality risks. The cost of the regulation is borne entirely by consumers. Some of the people who bear the costs of the regulation are poor; some are wealthy. For poor consumers, the average VSL is $1 million. For wealthy consumers, the average VSL is $20 million. The U.S. government uses a population-wide average, which is $12 million.

Use of the population-wide average is not in the interest of poor people. It is as if they are subjected to a forced transfer on terms that they dislike. They are effectively required to purchase more safety than they want. If their VSL is treated as if it were $12 million, they cannot gain; they can only lose. At least this is so if they do not suffer from a lack of information or some kind of behavioral bias.

A similar analysis would apply to the risks associated with greenhouse gas emissions. Suppose that emissions reductions reduce those risks. Suppose that poor people are being asked to pay more than they are willing to pay for the reduction. Here again, poor people are effectively required to purchase more safety than they want. That is not in their interest.

A notation: If we use a population-wide average in such circumstances, and add a distributional weight to the benefits enjoyed by poor people, we will make things even worse and possibly much worse.[15] Suppose that poor people are willing to pay two dollars to avoid a risk of 1 in 100,000. If we treat eliminating that risk as if it were worth far more to poor people than it is, by adding a distributional weight, we will be forcing poor people to pay much more than they want. That is not a very nice thing to do.

## PROBLEM 3

The U.S. government is regulating certain air pollutants, including greenhouse gases, to diminish risks, including mortality risks. Some of the people who bear the costs of the regulation are poor; most are wealthy. For poor people, the average VSL is $1 million. For wealthy people, the average VSL is $20 million. The U.S. government uses a population-wide average, which is $12 million.

Use of the population-wide average might well be in the interest of poor people; we cannot know without obtaining more information about the incidence of the costs. It is true that poor people are not willing to pay much to avoid a risk of 1 in 100,000. But it is not at all clear that they are being required to purchase more safety than they want. Because most of the people who are paying to reduce the risk are wealthy, poor people might be required to pay very little to avoid a risk of 1 in 100,000. If so, they will probably gain.

These conclusions are tentative, because poor people might turn out to be net losers from the regulation if they do not enjoy most of the benefits. Suppose that the benefits are concentrated among rich people so that they enjoy the overwhelming majority of the mortality and morbidity gains. We would have to obtain a host of numbers to know, but it is possible that poor people are net losers, even with a VSL of $10 million. Assume, for example, that they pay 49 percent of the total costs (which, let us suppose, are $800 million), but receive 2 percent of the total benefits (which are, let us suppose, $1 billion). If so, they are likely to be losers. We could imagine other assumptions that would lead to the opposite conclusion.

## PROBLEM 4

The U.S. government is regulating to protect the safety of construction workers in a well-defined region of the country; the benefits will be enjoyed entirely by those workers. The cost of the regulation would be passed onto relatively wealthy people. For the construction workers, the average VSL is $2 million. For those who bear the costs, the average VSL is $15 million. The U.S. government uses a population-wide average, which is $10 million.

Use of the population-wide average is likely to be in the interest of the construction workers. It is true that they are willing to pay only ten dollars to avoid a risk of 1 in 100,000. But it is not at all clear that they are being required to purchase more safety than that want. Because the people who are paying to reduce the risk are relatively wealthy, construction workers might be required to pay less than ten dollars—actually, zero dollars—to avoid a risk of 1 in 100,000. If so, they will gain.[16]

## PROBLEM 5

The U.S. government is regulating air pollution to diminish risks, including mortality risks. Most of the people who bear the costs of the regulation are poor; some are wealthy. For poor people, the average VSL is $1 million. For wealthy people, the average VSL is $20 million. The U.S. government uses a population-wide average, which is $10 million.

Use of the population-wide average is unlikely to be in the interest of poor people. They are willing to pay only ten dollars to avoid a risk of 1 in 100,000, and it appears quite possible that they are being required to purchase more safety than they want. Because most of the people who are paying to reduce the risk are poor, poor people might well be required to pay more than ten dollars to avoid a risk of 1 in 100,000. If so, they will lose.

As in problem 3, these conclusions are tentative, because we do not yet know about the incidence of the benefits. Suppose that the benefits are concentrated among poor people, so that they enjoy the overwhelming majority of the mortality and morbidity gains. We would have to obtain a host of numbers to know, but it is possible that poor people are net gainers, even with a VSL of $10 million. Assume, for example, that they pay 51 percent of the total cost (which is, let us suppose, $100 million), but receive 98 percent of the total benefit (which is, let us suppose, $2 billion). If so, they are likely to be gainers. We could imagine other assumptions that would lead us to conclude that poor people are net losers.

## WELFARE

My focus has been on whether the use of a uniform VSL is good for poor people. We are especially focusing on whether use of such a VSL is good for poor people in the context of climate policy. Turn now to the EPA's proposal to adjust VSL for per capita wealth so that the social cost of carbon depends on how wealthy the nation is. How should we think about that decision?

Chapter 1 argued in favor of the use of the global figure for the social cost of carbon and hence for the decision to consider the adverse effects of U.S. greenhouse emissions on other nations. If the goal is to turn those adverse effects into monetary equivalents, the use of the wealth-dependent VSL, in the relevant countries, is certainly a reasonable choice. A higher VSL—say, the U.S. VSL—would hardly hurt people in those countries. Actually, it would help them. But because it would give them an amount that outruns the VSL of their citizens, it would be a major subsidy, in excess of the monetary value of the risk.

It is true that use of a high VSL in the case of subsidies could be good for poor people but bad for those who pay for the subsidies. If so, what should be done? One question is how good it is, exactly, for poor people, and how bad it is, exactly, for everyone else. Another question is the size of the relevant populations. How many people are poor, and how many people are not poor? If a policy makes life a great deal better for a large number of poor people, and a tiny bit worse for a small number of rich people, it would seem to be an excellent idea.[17] If a policy makes life a tiny bit better for a small number of poor people, and a great deal worse for a large number of rich people, it would not seem to be an excellent idea.

Harder cases would involve different magnitudes of effects and different sizes of populations. Suppose, for example, that a policy made life non-trivially better for a significant number of poor people, but also made life nontrivially worse for a significant number of people who are not poor. We would need to have more details to know whether the policy is producing net welfare gains.

Purely on welfare grounds, we should now have something like a general framework with which to approach a host of climate-related problems. There is a difference between funding programs and regulatory programs. Subsidies will help those being subsidized, and the more, the better. Regulations may not help their intended beneficiaries. One implication is that regulations that increase health and safety, or that promote environmental protection (including greenhouse gas emissions reductions), may or may not be more appealing if we include *distributional weights*. A uniform VSL can be seen as a kind of distributional weight, and it may or may not help people, depending on the incidence of costs and benefits. Another implication is that a uniform VSL, combined with distributional weights for benefits that accrue to poor people, might seem appealing on normative grounds, but that it might produce terrible outcomes for poor people if they end up footing the bill.

There are also questions about distribution as such. Suppose (1) that there is a net loss in welfare (which is clearly bad) and (2) that poor people gain a great deal and people who are not poor lose slightly more than poor people gain. We might favor that policy on grounds of distributional justice. To know whether we should, we should want to know the magnitude of the loss in welfare, the magnitude of the gain to poor people, and the magnitude of the loss to people who are not poor. And if we are prioritarians, we will give special attention to those at the bottom of the ladder, and accept overall losses if they are the price we have to pay in order to help that group.[18]

# 5 ADAPTATION

In recent years, worldwide losses from natural disasters have increased dramatically.[1] The sources of the losses vary, but in some places, wildfires are a significant problem. Over a five-year period, California alone experienced over $40 billion in wildfire-related losses.[2] It might be tempting to think that because of climate change and other contributors to the relevant risks, these kinds of losses are essentially inevitable. To be sure, reduction of greenhouse gas emissions can slow their growth, and hence much of the focus of climate law and policy, and much of the focus of this book, has been on such reductions. As they say, less is more, and in this context, less is better.

But consider the following evidence. California has adopted a series of wildfire standards, usually prompted by horrific events.[3] These standards impose an assortment of requirements. For example, they require fire-resistant roofing or maintenance of vegetation near the home. They require decks and building appendages to consist of noncombustible materials. They require vents to be covered by a fire-resistant mesh. They require exterior siding to be fire-resistant.

Although California's standards are frequently mandatory, they vary over both time and space. For purposes of empirical investigation, at least, that is a big advantage; it makes it possible to measure the effects of mandatory codes. What do they do? Are they a good idea?

Patrick Baylis and Judson Boomhower have analyzed these questions in detail.[4] They find that even during a catastrophic wildfire, *more than half of exposed homes end up surviving*. They also find that mandatory codes make a major difference in the probability of survival. If a 1990 home and a 2008 (or later) home experience an identical wildfire, the 1990 home, not governed by a code, is about 40 percent less likely to survive than the home from 2008 or later, which is governed by a code. Compliance with a code also benefits one's neighbors. If a neighboring home is less than ten meters away, the likelihood that one's home will be destroyed by a fire is 6 percent lower if the neighboring home is code-compliant. It is true that code compliance can be costly. But in California's most fire-prone areas, the benefits of wildfire building codes unambiguously exceed the costs.

Here, then, is a simple but significant lesson about the potential benefits of policies and reforms that reduce climate-related risks through adaptation. This lesson is only one of many with respect to the benefits from efforts at adaptation.[5]

It is customary to divide climate policy into three categories: mitigation, adaptation, and resilience.[6] As noted, *mitigation* refers to reductions in greenhouse gas emissions. The difference between adaptation and resilience is less simple.[7] *Resilience* is sometimes taken to refer to the capacity to absorb and respond to the effects of potentially hazardous events.[8] *Adaptation* is sometimes taken to refer to adjustments to expected effects or risks, so that harms are moderated.[9] Because the two terms overlap, I shall use *adaptation* to cover all efforts to reduce the adverse effects of climate-related risks so as to ensure that to the extent possible, a hotter world, or a world with a more volatile climate, is not a more dangerous world.

We could easily imagine, for example, a California or Oregon five years hence that has done relatively little to adapt to the risks of wildfire, extreme heat, hurricanes, and flooding. By contrast, we could imagine a California or Oregon that has done a great deal to adapt to those risks so that they impose much less in the way of harm. If every state in the United States adopted wildfire codes akin to those in California, it is clear that the harms of wildfire would be significantly reduced. At the same time, it is not at all clear that every state in the United States should adopt such codes—a point to which I will return.

Of course, the lessons for the United States extend far more broadly. Adaptation is essential for nations all over the world. It is crucial for India, China, Pakistan, the Central African Republic, and Italy. But how much adaptation? And what kind of adaptation?

It is important to note that policymakers have a toolbox for promoting adaptation (just as they have a toolbox for promoting mitigation). We have seen that they might create incentives—for example, by giving out money, goods, or technical assistance to encourage private and public institutions to adapt.[10] We have also seen that they might impose regulation—for example, by following California's example and requiring various steps to reduce wildfire-related risks. (Those steps might or might not be subsidized.) They might engage in nudging—for example, by providing information about climate-related risks (educative nudges) and how to reduce them,[11] by making climate-related risks more salient,[12] by making it easier to take steps to reduce such risks,[13] or by making enrollment in risk-reducing actions essentially automatic.[14]

Each of these steps will have both costs and benefits. We could imagine costly adaptation efforts that would do a great deal of good. We could imagine costly adaptation efforts that would achieve very little. A large economic subsidy might not produce much in terms of benefits. A regulatory mandate could cost a great deal but deliver nothing. Nudges are typically inexpensive, and in part for that reason have been found to be highly cost-effective,[15] especially when they involve architectural interventions, which alter the underlying design of social environments (as, for example, through default rules).[16]

*Nudging climate adaptation* would be an eminently worthwhile effort. But by itself, information disclosure may or may not deliver much in the way of adaptation; everything depends on the context. Inertia or optimistic bias might ensure that disclosure is ineffective.

All of the tools for adaptation will also have distributive effects. As we saw in chapter 4, subsidies are relatively straightforward. People benefit from subsidies! Payments to poor people, or to people in poor nations, are highly likely to help. As we also saw, regulation is different; whether it will help, and how it will help, depends on the incidence of benefits and costs. Information may or may not be especially helpful for those at the bottom. Under imaginable circumstances, nudging will indeed benefit those at the bottom of economic ladder.[17] But it might not do that at all.

We have seen that much of public policy has focused on the question of mitigation. We have also seen that many people think it important to set some kind of ceiling on increases in global temperatures—perhaps 2°C, perhaps 1.5°C[18]—and to achieve emissions reductions that will ensure that the ceiling is not exceeded.[19] Recall that approaches of this kind raise an assortment of questions. Where, exactly, does the stated ceiling come from? Is it based on an assessment of costs and benefits, such that 1.5°C, and no more and no less, is clearly known to be optimal? Is it based on a judgment that once a ceiling is exceeded, the harms from climate change will be truly catastrophic, and that within the ceiling, the harms would be tolerable?[20]

If so, the argument for a stated ceiling would be easy to understand. Suppose that you are told that if you eat five sausages, you will be fine, but that if you eat a sixth, you will die; if so, you had better not eat the sixth, even if you really love sausages. But in light of the multiple uncertainties associated with climate change and the relationship between emissions and harms, it would be surprising if anyone could say, with real confidence, that any particular ceiling on emissions, or on anticipated warming, is the right one.[21]

Nothing here is meant to raise questions about aggressive efforts to reduce greenhouse gas emissions. For various reasons, including the risk of catastrophe, those efforts are essential.[22] At the same time, reasonable people have raised serious questions about the feasibility of meeting some prominent current goals, including a 1.5°C ceiling and even a 2°C ceiling.[23] This is a domain in which predictions are especially hazardous, but according to some accounts, it is extremely unlikely that nations will be able to do what must be done to meet those goals. As Robert Pindyck puts it:[24]

> The sad but fundamental problem is that over the coming decades worldwide GHG emissions are likely to grow, and atmospheric GHG concentrations will surely grow. This will be the case under any conceivable (but realistic) climate policy. Current targets for GHG emission reductions vary, but . . . even the more aggressive targets are insufficient to prevent increases in atmospheric GHG concentrations. (As part of the Paris Agreement, for example, China pledged to reduce the *growth rate* of its emissions between now and 2030, but it did not pledge to reduce its *level* of emissions.) Furthermore, it is very unlikely that the world will even come close to meeting current targets for emission reductions. Thus we must come to grips with the likelihood of a global mean temperature increase over the next 50 to 70 years that could turn out to be 3°C or even higher—well above the 1.5°C to 2°C that many climate scientists and policy analysts have argued is a critical limit. This could lead to rising sea levels, greater variability of weather, more intense storms, and other forms of climate change.

Perhaps this view is too pessimistic; perhaps political will, technological innovation, or both will lead to less warming than many people anticipate. But in light of the multiple harms likely to arise at 1.5°C or 2°C, let alone 3°C or higher, it makes sense to focus intensely on adaptation.

## THE ETHICS OF ADAPTATION

But what kinds of adaptation? What kind of framework might be useful?[25] The study of wildfire standards in California offers important clues. The study focuses on the state's most fire-prone areas, where the benefits of those standards are clearly worth the costs. In areas that are least fire-prone, their costs are not necessarily justified.[26] The point should be intuitive: precautions against low-probability risks might not be worthwhile, depending on their cost, on how low the probability is, and on the magnitude of the bad outcome if it should come to fruition. Consider the planned seawall designed to encircle southern Manhattan, with the goal of preventing flooding from storm. The federal government allocated $176 million for the project in 2016, which seems like a large sum. But subsequent analysis produced a plan with an expected cost of $1 billion. And by 2020, the estimated cost rose to $119 billion, at which point the whole project was put on hold.

A competent cost-benefit analysis should take all of the relevant considerations on board, and that analysis is a valuable clue to the welfare effects of adaptation measures. Indeed, the cost-benefit analysis of adaptation may well be far more tractable than that of mitigation. We do not have to rely on integrated assessment models, or to pick among them, to know whether it makes sense to take specific precautions with respect to flooding, extreme heat, and wildfires.

Still, we have seen enough to see that there are problems. The first involves valuation. Assume that there are two communities in California: Rich and Poor. In Rich, the median value of a home, vulnerable to wildfire, is $5 million. In Poor, the median value of a home, vulnerable to wildfire, is $300,000. Rich and Poor are equally vulnerable to wildfire. A subsidy to Rich, designed to fund measures to reduce the risks associated with wildfire, will produce far higher benefits (and net benefits) than an equivalent subsidy to Poor for that purpose. Does it follow that on welfare grounds, Rich should receive the subsidy, and that perhaps Poor should not?

Not at all! Notwithstanding the monetary figures, the citizens of Poor might receive the same *welfare* from their homes as the citizens of Rich (or perhaps more). A policy that subsidizes Rich but not Poor might well be mistaken to rely on monetized benefits, because these do not adequately capture the welfare effects of subsidies. Recall: Does a wealthy person get more welfare from a $5 million home than a poor person gets from a $300,000 home? That is not at all clear. The opposite might be the case.

The second problem involves distributional equity. Suppose that, in fact, the citizens of Rich would receive more welfare from a given subsidy than would the citizens of Poor. Does it follow that officials should subsidize the citizens of Rich? Not at all.

As we have seen, prioritarianism suggests that we should devote special attention to the welfare of those who are least well-off.[27] Imagine that the world consists of two people, Mary and Edna. Mary has 100 units of welfare; Edna has 1 unit of welfare. If we choose intervention A, both will gain 20 units of welfare, so that Mary will have 120, and Edna will have 21. If we choose intervention B, Mary will gain 10 units of welfare and Edna will gain 28, so that Mary will have 110, and Edna will have 29. Though intervention A results in more aggregate welfare (141 is larger than 139), prioritarians would prefer intervention B, not because it results in a more *equal* distribution, but because it *gives more help to the person at the bottom*. As for individuals, so for groups: we might give priority to those whose welfare is lowest and sacrifice aggregate welfare in order to achieve that goal.[28]

But how much should we sacrifice? The answer must depend on the right specification of prioritarianism. A situation in which Mary has 120 and Edna has 5 might be better than one in which Mary has 50 and Edna has 6 (or not). A situation in which Mary has 50 and Edna has 6 might be better than one in which Mary has 12 and Edna has 8 (or not).

Prioritarianism has strong defenders, though it also raises many questions, especially if it is used for regulatory purposes. One set of questions involves *incidence*: a fuel economy rule will have an assortment of effects on consumers, workers, and investors, and even if one set of effects is especially beneficial to the least well-off (say, less pollution), another set of effects might be especially harmful to them (say, increased prices and fewer jobs). An occupational safety regulation might reduce risks to poor workers, but it might also result in decreases in their wages and in reduced working hours. Another set of questions cuts deeper. We might want to know why, exactly, Edna has 1 while Mary has 100. Did Mary work harder? Did she have better luck? Is she smarter or stronger? We might also want to know the dynamic effects of prioritarianism. If we are to prioritize the welfare of the least well-off, will we reduce effort and growth?

Rawls's Difference Principle allows inequalities to the extent that they benefit the least well-off. Thus: "Social and economic inequalities are to be arranged so that they are both (a) to the greatest expected benefit of the least advantaged and (b) attached to offices and positions open to all under conditions of fair equality of opportunity."[29] The Difference Principle of course raises many questions, and it is not at all clear how to apply the Difference Principle when making policy choices in the context of climate change; Rawls's own focus was on the basic structure of society, not on particular interventions. For present purposes, the only point is that prioritarianism might misfire if its use ends up having harmful aggregate effects, so that everyone ends up with 1, or perhaps 0.5.

A third problem actually consists of two separate ones: (1) very low-probability risks and (2) uncertainty. Suppose that in a community, the annual risk of an extreme climate-related event is 1 in 100. Suppose that if the event occurred, the amount of damages would be $100 million. Would it make sense to spend $1 million, annually, to eliminate that risk? Might it make sense to spend more? People buy insurance; perhaps a degree of risk aversion is justified in these circumstances. But how much risk aversion? Or suppose that in a community, the annual risk of an extreme climate-related event is 1 in X, where X is unknown. Suppose that if the event occurred, the lower bound of the relevant amount of damages would be $100, and the upper bound is unknown. What then?

Consider in this regard a passage from John Maynard Keynes, who lived through the Great Depression and World War II, who spent much of young adulthood in same-sex relationships before he fell head over heels in love with a woman, and who knew a great deal about the unforeseeable:

> By "uncertain" knowledge, let me explain, I do not mean merely to distinguish what is known for certain from what is only probable. The game of roulette is not subject, in this sense, to uncertainty; nor is the prospect of a Victory bond being drawn. Or, again, the expectation of life is only slightly uncertain. Even the weather is only moderately uncertain. The sense in which I am using the term is that in which the prospect of a European war is uncertain, or the price of copper and the rate of interest twenty years hence, or the obsolescence of a new

> invention, or the position of private wealth-owners in the social system in 1970. About these matters there is no scientific basis on which to form any calculable probability whatever. We simply do not know.[30]

With respect to some climate-related risks, communities might be in the vicinity of this problem, sometimes called *Knightian uncertainty*.[31] As Pindyck states, "No one could have predicted the arrival and strength of Hurricane Katrina in August 2005, never mind its impact on New Orleans. Furthermore, climate change means that the statistics for storm surges that might have been valid in 2005 or 2020 are unlikely to be valid two or three decades from now."[32] If so, communities might want to take steps to prevent the worst-case scenarios. But if those steps are very expensive, they might want to hesitate.

In the relatively short term, it should be possible to match particular climate-related risks to particular strategies for adaptation, with some clarity on the welfare effects of those strategies and their distributional impacts.[33] There are two essential steps. The first is to give communities all over the world far more clarity about the risks they face and how to combat them. Consider https://www.heat.gov, a U.S. government website designed to do exactly that with respect to extreme heat.[34] The second step is to give communities all over the world the resources they need, alongside technical assistance. Consider Building Resilient Infrastructure and Communities (BRIC), a U.S. government program designed to do exactly that with respect to climate-related risks.[35] Even as we explore the best approaches to the problem of mitigation, we need to explore, with real urgency, how best to protect vulnerable people and places from what is now here, and likely to get much worse. As we have seen, wealthy countries are morally obliged to do a great deal to fund that protection.

# 6 CONSUMERS

Climate change is in large part a product of consumer choices—of what people buy, whether we are speaking of motor vehicles, energy, or food. In this chapter, I am going to be exploring what might be done to improve those choices, with reference to the well-being of consumers themselves, and also with reference to the reduction of greenhouse gas emissions. To do that, I am going to venture far and wide, with some issues and problems that do not involve climate change at all. We will not deal only with ethics, not by any means, but we will get to ethics, and soon enough.

Let us begin with three sets of findings:

1. On average, people appear to benefit from home energy reports; they save money, on average, and they are willing to pay, on average, a positive amount to receive such reports. But some people are willing to pay far more than others.[1] In fact, some people are willing to pay *not* to receive home energy reports. They believe that they would be better off without them. While home energy reports are designed to save consumers money and to reduce externalities, sending such reports to some people seems to have costs in excess of benefits. More targeted use of home energy reports could produce significant welfare gains.
2. On average, labels on sugary drinks affect consumer behavior, and in what seems to be the right way; such labels reduce demand for such drinks. But labels on sugary drinks have greater effects on some consumers than on others.[2] Disturbingly, labels can lead people who do not have self-control problems to consume less in the way of sugary drinks while

having a significantly smaller effect on people who do have self-control problems. In addition, many people do not like seeing graphic warning labels. The average person in a large sample reported being willing to pay about one dollar to *avoid* seeing graphic warning labels. It is likely that such labels are helping some and hurting others. It is possible that such labels are on balance causing harm.

3. There is evidence that calorie labels have welfare benefits.[3] At the same time, they seem to have a greater effect on people who lack self-control problems than on people who suffer from such problems. It is possible that in some populations, calorie labels affect people who do not need help, and have little or no effect on people who do need help, except to make them feel sad and ashamed.

From these sets of findings, we can draw three simple conclusions, all of which have implications for climate policy and associated ethical issues. *First,* interventions may have either positive or negative *hedonic* effects. People might like seeing labels, or they might dislike seeing labels. *Second,* interventions might well have different effects on different populations. Under favorable conditions, they might have large positive effects on a group that needs help, and small or no effects on a group that does not need help. Under unfavorable conditions, they might have small or no effects on a group that needs help, and large effects on a group that does not need help. Large effects on a group that does not need help may not much improve that group's welfare. For example, people who have no need to change their spending patterns, or their diets, might end up doing so. *Third,* and consistent with the second conclusion, an understanding of the average treatment effect does not tell us what we need to know.[4] We might have a large effect, on average, and even so, we might not know whether the label is improving social welfare.

These points about labels can be made about a wide range of interventions, emphatically including those that involve climate change. They hold for automatic enrollment. It is possible that automatic enrollment in some plan (say, green energy) will have no effects on people who benefit from enrollment (because they would enroll in any case) while harming people who do not benefit from enrollment (some or many who lose may not opt out, perhaps because of inertia). They hold for taxes. It is possible that taxes will have little or no effect on the people they are particularly intended to help while having a significant adverse effect on people they are not (particularly) intended to help. They hold for mandates and bans. A ban on some activity or product might, on balance, hurt people who greatly benefit from that activity or product while helping people who lose only modestly from it. (Consider bans on the purchase of incandescent lightbulbs.) In all of these cases, more targeted action would be better than "mass" action.

## CHOICE ENGINES

For retirement plans, many employers use something like a *Choice Engine*.[5] Let us understand the term to refer to an online technology that receives a few, or perhaps more than a few, details about choosers, and that responds by offering a clear, simple set of options, or perhaps a "default" option, which would be accompanied by a right to opt out and to choose alternatives that are compared with the default along specific dimensions.

Consider the domain of savings. Employers know a few things about their employees (and possibly more than a few). On the basis of what they know, they might automatically enroll their employees in a specific plan. In practice, the plan is frequently a diversified, passively managed index fund. Employees can opt out and choose a different plan if they like.

Alternatively, employers might offer employees a specified set of options, with the understanding that all of them are suitable, or suitable enough. (Options that are not suitable are not included.) They might provide employees with simple information to help them to choose among them. The options might be identified or rethought with the assistance of artificial intelligence or some kind of algorithm. Here is one reasonable approach: automatically enroll employees in a plan that is most likely to be improve their well-being, given everything relevant that is known about them. Identification of that plan might prove daunting, but a large number of plans can at least be ruled out. Note that if the focus is on improving employee well-being, we are not necessarily speaking of revealed preferences.

For retirement savings, we can easily imagine many different kinds of Choice Engines. Some of them might be mischievous; some of them might be fiendish; some of them might be random; some of them might be coarse or clueless; some of them might show behavioral or other biases of their own; some of them might be self-serving. For example, people might be automatically enrolled in plans with high fees. They might be automatically enrolled in plans that are not diversified. They might be automatically enrolled in money market accounts. They might be automatically enrolled in dominated plans. They might be automatically enrolled in plans that are especially ill-suited to their situations. They might be given a large number of options and asked to choose among them, with little relevant information, or with information that leads them to poor choices.

The general point is that in principle, Choice Engines might work to overcome behavioral biases, including those related to climate policy.[6] For retirement plans, Choice Engines may or may not be paternalistic. If they are not paternalistic, it might be because they simply provide a menu of options, given what they know about relevant choosers. If they are paternalistic, they might be mildly paternalistic, moderately paternalistic, or highly paternalistic. A moderately paternalistic Choice Engine might impose nontrivial barriers to those who seek certain kinds of plans (such as those with high fees). The barriers might take the form of information provision, "Are you sure you want to?" queries, and requirements of multiple clicks. We might think of a moderately paternalistic Choice Engine as offering *light patterns*, as contrasted with *dark patterns*.[7] A highly paternalistic Choice Engine might forbid employees from selecting any plan other than the one that it deems in the interest of choosers, or might make it exceedingly difficult for employees to do that.

Choice Engines of this kind might be used for any number of choices, including (to take some random examples) choices of food, dogs, laptops, mystery novels, cellphones, shavers, shoes, tennis racquets, and ties. Choice Engines may or may not use AI, and if they do, they can use AI of different kinds. Consider this question: What kind of car would you like to buy? Would you like to buy a fuel-efficient car that would cost you $800 more up front than the alternative, but that would save you $8,000 over the next ten years? Would you like to buy an energy-efficient refrigerator that would cost you $X today, but save you ten times $X over the next ten years? What characteristics of a car, or a refrigerator, matter most to you? Do you need a large car? Do you like hybrids? Are you excited about electric cars, or not so much?

A great deal of work finds that consumers suffer from *present bias*.[8] Current costs and benefits loom large; future costs and benefits do not. For many of us, the short term is what most matters, and the long term is a foreign country. The future is Laterland, a country that we are not sure that we will ever visit. This is so with respect to choices that involve money, health, safety, the environment, and more.

AI need not suffer from present bias.[9] Imagine that you are able and willing to consult AI to ask it what kind of car you should buy. Imagine too that you discover that you are, or might be, present-biased, in the sense that you prefer a car that is not (according to AI) the one that you should get. What then? We could easily imagine Choice Engines for motor vehicle purchases in which different consumers provide relevant information about their practices, their preferences, and their values, and in which the relevant Choice Engine immediately provides a set of options—say, *good*, *better*, and *best*. Something like this could happen in minutes or even seconds, perhaps a second or two. If there are three options—good, better, and best—then verbal descriptions might explain the ranking. Or a Choice Engine might simply say: *best for you*. It might do so while allowing you to see other options if you indicate that you wish to do so. It may or may not be paternalistic, or come with guardrails designed to protect consumers against serious mistakes.[10]

## CLIMATE CHANGE AND CONSUMPTION: INTERNALITIES AND EXTERNALITIES

Attempting to respond to the kinds of findings with which I began, Choice Engines might well focus solely on particular consumers and what best fits their particular situations. They might ask, for example, about what particular consumers like most in cars, and they might take account of the full range of economic costs, including the costs of operating a vehicle over time. If so, Choice Engines might have a paternalistic feature insofar as they suggest that car A is "best" for a particular consumer, even if that consumer would not give serious consideration to car A. A Choice Engine would attempt to overcome both informational deficits and behavioral biases on the part of those who use them. Freedom of choice would be preserved, in recognition of the diversity of individual tastes, including preferences and values.

Present bias is, of course, just one reason that consumers might not make the right decisions, where *right* is understood by reference to their own welfare. Consumers might also suffer from a simple absence of information, from status quo bias, from limited attention, or from unrealistic optimism. If people are making their own lives worse for any of these reasons, Choice Engines might help. They might be paternalistic insofar as they respond to behavioral biases on the part of choosers, perhaps by offering recommendations or defaults, perhaps by imposing various barriers to choices that, according to the relevant Choice Engine, would not be in the interest of choosers.

These points involve ethical issues, though not the ethical issues that are my main focus in this book. Is paternalism justified? Is it justified if it preserves freedom of choice? In my view, the answer to the second question, at least, is an enthusiastic "yes." (The answer to the first question is a firm "sometimes.") But a defense of such conclusions would take many pages.[11] The principal issue here involves not paternalism but externalities: the harms that people impose on others. A major goal of Choice Engines might be to address that problem. Focusing on greenhouse gas emissions, for example, Choice Engines might use the social cost of carbon to inform choices. Suppose, for simplicity, that it is one hundred dollars. Choice Engines might select good, better, and best options, incorporating that number. A Choice Engine that includes externalities might do so by default (my preference), or it might do so if and only if choosers explicitly request it to do so (not bad).

Choice Engines might be designed in different ways. They might allow consumers to say what they care about, including externalities. They might be designed so as to allow consumers to see good, better, and best with and without externalities. They might be designed so as to allow a great deal of transparency with respect to when costs would be incurred. If, for example, a car would cost significantly more up front, but significantly less over a period of five years, a Choice Engine could reveal that fact.

We could imagine a *keep it simple* version of a Choice Engine, offering only a little information and a few options to consumers. We could imagine a *tell me everything* version of a Choice Engine, living up to its name. Consumers might be asked to choose what kind of Choice Engine they want. Alternatively, they might be defaulted into keep it simple or tell me everything, depending on what AI thinks they would choose, if they were to make an informed choice, free from behavioral biases.

## CHASTITY, TOMORROW

Let us now focus in particular on *impatience*, which is of course a capacious idea, a kind of umbrella concept that includes diverse psychological phenomena, some involving cognition and others involving emotions. The term might refer to a sense of *urgency*: a particular problem or goal needs immediate attention. If people have a physical need or are feeling intense but unrequited romantic love, the situation might seem to require a solution now, not later. In government, a public official might be impatient to implement a policy that would immediately improve the lives of millions of people, even though the policy might be better if people worked on it for a longer time. In business, a construction contractor might be impatient to finish a building, even though the building might look better, and be better, if he took his time. In a university, a professor might be impatient to publish a new article, even though the article would much benefit from a few more months, or perhaps years, of research and thought.

Impatience typically refers to an inability to wait, whether or not waiting is a good idea. The idea of the *fierce urgency of now*, emphasized by Barack Obama in the 2008 presidential campaign, captures a psychological state that may or may not be fully rational. There is a risk that it might induce current action when people would be better off if they bided their time. But even if it is not fully rational, impatience might provide a necessary impetus toward action that would not otherwise occur and that is immensely beneficial—smoking cessation, loss of weight, a breakup of a less-than-good relationship, the change of a job, relocation, reform of the energy system, completion of a project. The *planning fallacy* creates serious problems for managers of all kinds, and impatient managers can make the planning fallacy into less of a fallacy. Having the patience of Job can calm the mind in the most turbulent waters (say, when recovering from an illness or a terrible setback), but it can also ensure that nothing gets ever done.

In short, patience has costs, and they might be unreasonably high. There is a saying: "Enjoy life now. This is not a rehearsal." St. Augustine said: "God give me chastity—tomorrow." If you are too patient, you might lose out on a world of good things. Someone else might take them (or you might die first). What is needed is an optimal level of patience. One advantage of the term *impatience* is that it unambiguously connotes *insufficient* patience, even if we also note that it might be a valuable or essential spur. Present bias can be seen as a major source of impatience, or as one of its many faces.

Present bias is typically taken to involve an indefensibly high discount rate or hyperbolic discounting. To that extent, it represents a form of bounded rationality, broadly associated with the work of Herbert Simon, but growing more specifically out of recent work on systematic departures from perfect rationality. Impatience as such may or may not be fully rational, or a source of welfare losses, but if people impose high costs on their future selves, they might be damaging or even ruining (much of) their lives. In many domains, consumers show present bias. A large-scale study finds that after a significant correction of an erroneously stated miles per gallon (MPG) measure, consumers were relatively unresponsive; they did not make different choices.[12] As Gillingham et al. write, "Using the implied changes in willingness-to-pay, we find that consumers act myopically: consumers are indifferent between $1.00 in discounted fuel costs and $0.16–0.39 in the purchase price when discounting at 4 percent."[13] That is present bias in action. It hurts consumers, and it is bad for the environment.

What, if anything, should be done in response? Can Choice Engines help? Can AI? Recall the problem of heterogeneity; recall too the difference between internalities and externalities.

## MISTAKES, WELFARE, AND LABELS

Most motor vehicles emit pollution, including greenhouse gases, and the use of gasoline increases national dependence on foreign oil. On standard economic grounds, the result is a market failure in the form of excessive pollution, and some kind of cap-and-trade system or corrective tax is the best response, designed to ensure that drivers internalize the social costs of their activity. The choice between cap-and-trade programs and carbon taxes raises a host of important questions. But the more fundamental point is that economic incentives of some kind, and not mandates, are the appropriate instrument. Simply put, incentives are far more efficient; for any given reduction in pollution levels, they impose a lower cost.

For obvious reasons, a great deal of recent analysis has been focused on greenhouse gas emissions and how best to reduce them. In principle, regulators have a host of options. They might create subsidies—say, for electric cars. They might use nudges—say, by providing information about greenhouse gas emissions on fuel economy labels. They might impose regulatory mandates—say, with fuel economy and energy efficiency standards. Careful analysis suggests that carbon taxes can produce reductions in greenhouse gas emissions at a small fraction of the cost of fuel economy mandates.[14] By one account, "a fuel economy standard is shown to be at least six to fourteen times less cost effective than a price instrument (fuel tax) when targeting an identical reduction in cumulative gasoline use."[15]

These are points about how best to reduce externalities. But behaviorally informed regulators focus on consumer welfare, not only externalities. They are concerned about a different kind of market failure, one that is distinctly behavioral. Regulators speculate that at the time of purchase, many consumers, focused on the short term, might not give sufficient attention to the full costs of driving a car. Even if they try, they might not have a sufficient understanding of those costs, because it is not simple to translate differences in MPG into economic and environmental consequences.[16] An obvious response, preserving freedom of choice, would be disclosure, in the form of a fuel economy label that would correct that kind of behavioral market failure, and in a sense give people some visibility into Laterland.

Such a label might, for example, draw consumers' attention to the long-term costs or savings associated with motor vehicles. And indeed, the existing label in the United States does exactly that; it specifies annual fuel costs and also five-year costs (or savings) compared to the average vehicle. Figure 6.1 gives one example.

In principle, such a label, if designed to counteract present bias, should solve the problem. In short: labels should be used to promote consumer welfare, by increasing the likelihood that consumers will make optimal choices, and corrective taxes should be used to respond to externalities. A label protects consumers from their own mistakes, in terms of their own self-interest; corrective taxes protect those who are injured by pollution.

But labels are not Choice Engines. In the face of present bias, it would be possible to wonder whether a label will be sufficiently effective. It might not be easy to get people to attend to Laterland. This is an empirical question, not resolvable in the abstract. Perhaps some or many consumers will pay too little attention to the label, and hence will not purchase cars that would save them a significant amount of money. If some or many consumers are genuinely inattentive to the costs of operating a vehicle at the time of purchase, and if they do not make a fully informed decision in spite of

FIGURE 6.1
Fuel economy label

adequate labeling (perhaps because of present bias), then it is possible to justify fuel economy standards with a level of stringency that would be difficult to defend on standard economic grounds. (We will return to Choice Engines in due course; I am going to focus, in the next pages, on the question of whether such standards, so central to climate change policy, might have a behavioral justification.)

In support of that argument, it would be useful to focus directly on two kinds of consumer savings from fuel economy standards, involving internalities rather than externalities: money and time. In fact, the vast majority of the quantified benefits from recent fuel economy standards from multiple administrations have been said to come not from environmental improvements, but from money saved at the pump; turned into monetary equivalents, the time savings are also significant. Under the Obama administration, the Department of Transportation found consumer savings of about $529 billion, time savings of $15 billion, energy security benefits of $25 billion, carbon dioxide emissions reductions benefits of $49 billion, other air pollution benefits of about $14 billion, and just under $1 billion from reduced fatalities (as a result of cleaner air).[17] The total projected benefits were $633 billion over fifteen years, of which a remarkable 84 percent come from savings at the pump, and no less than 86 percent from those savings along with time savings (because drivers do not have to go to the gas station so often).[18]

In its own rulemaking, the Trump administration rethought those numbers by reference to recent works[19] raising questions about whether consumers are insufficiently attentive to the economic savings, but it projected the consumer savings to be in the same general vicinity and actually even higher.[20] The Biden administration produced numbers that are broadly similar to those in predecessor administrations, in the sense that once again, the strong majority of the monetized benefits come from consumer savings.[21]

But these justifications run into two problems. The first is that of heterogeneity: Some consumers are alert to those savings, and others are not. Some people care a great deal about fuel economy, and others do not. How ought that problem to be handled? Might Choice Engines help?

The second problem is that on standard economic grounds, it is not at all clear that consumer benefits from money and time savings are entitled to count in the analysis, because they are purely private savings, and do not involve externalities in any way.[22] In deciding which cars to buy, consumers can certainly take account of the private savings from fuel-efficient cars; if they choose not to buy such cars, it might be because they do not value fuel efficiency as compared to other vehicle attributes, such as safety, aesthetics, and performance. Where is the market failure? If the problem lies in a lack of information, the standard economic prescription is the same as the behaviorally informed one: *Fix the label and provide that information so that consumers can easily understand it.* More simply, *make Laterland fully present.*

We have seen, however, that even with the best fuel economy label in the world, consumers might turn out to be insufficiently attentive to the benefit of improved fuel economy at the time of purchase, not because they have made a rational judgment that these benefits are outweighed by other factors, but simply because consumers focus on other, more current variables, such as performance, size, and cost. So the problem may be not one of information, but of present bias and insufficient attention. The behavioral hunch is that automobile purchasers do not give adequate consideration to economic savings. Apart from savings, there is the question of time: How many consumers think about time savings when they are deciding whether to buy a fuel-efficient vehicle? Consumers might benefit from Choice Engines.

## MORE WELFARE

If a person at time 1 pays little attention to his well-being (in terms of health, money, or otherwise) at time 2 or time 3, ought we to insist that he has departed from perfect rationality? That question immediately raises philosophical questions. If we see a person at various times as a series of selves, rather than as one self, the answer is not straightforward. John right now might think that John next year or the year after is a different person, and not worthy of the same attention as John right now. If so, John right now might not care so much about the suffering or deprivation of future Johns. The claim of present bias depends on a judgment that a person, extending over time, really is the same person, which means that indifference to or disregard for one's own welfare is a genuine mistake.

I will be accepting that judgment here, with the suggestion that rational agents should aggregate the well-being of their selves, extending over time. That suggestion is consistent with the proposition that some discount rate is appropriate, both because one might die (what is the probability that one will be alive twenty years?) and because money, at least, can be invested and made to grow (which means that a given amount of money is worth more today than next year). *Present bias* refers to indifference to one's own future welfare that cannot be explained by reference to a reasonable discount rate.

In light of the relevant findings, demonstrating the occasional human propensity to neglect the future, it is natural to ask whether mandates and bans have a fresh justification.[23] The motivation for that question is clear: If we know that people's choices lead them in the wrong direction, why should we insist on freedom of choice? In the face of human errors, is it not odd, or even perverse, to insist on that form of freedom? Is it not especially odd to do so if we know that in many contexts, people choose wrongly, thus injuring their future selves?

If people are suffering from present bias or a problem of self-control, and if the result is a serious welfare loss for those very people, there is an argument for some kind of public response, potentially including mandates. If, for example, people are present-biased, they might not protect their future selves. When people are running high risks of mortality or otherwise ruining their lives, it might make sense to adopt a mandate or a ban on welfare grounds. After all, people have to get prescriptions for certain kinds of medicines, and even in freedom-loving societies, people are forbidden from buying certain foods or running certain risks in the workplace, simply because the dangers are too high. Many occupational safety and health regulations must stand or fall on behavioral grounds; they forbid workers from voluntarily facing certain risks, perhaps because present bias might lead them to do so unwisely. We could certainly identify cases in which the best approach is a mandate or a ban, because that response is preferable, from the standpoint of social welfare, to any alternative, including economic incentives or defaults.

Many different areas might be chosen to explore that possibility. My aim here is to explore the possibility of defending fuel economy mandates, and also energy efficiency mandates, as opposed to economic incentives, by reference to present bias. The most general point is that such mandates may reduce *internalities*, understood as the costs that choosers impose on their future selves.[24] Fuel economy mandates might simultaneously reduce internalities and externalities. On plausible assumptions about the existence and magnitude of consumer errors (stemming from, for example, present bias), such mandates might turn out to have higher net benefits than carbon taxes, because the former, unlike the latter, deliver consumer savings. To say the least, this is not a conventional view, because fuel economy standards are a highly inefficient response to the externalities produced by motor vehicles, especially when compared to optimal corrective taxes.[25]

As we will see, everything turns on whether the plausible assumptions turn out to be true. My goal is not to run the numbers or to reach a final conclusion, but to make three more general points. The first is that that in light of behavioral findings about present bias, fuel economy mandates might be amply justified on welfare grounds. The second is that the standard economic preference for economic incentives over mandates misses something of considerable importance. In brief, it misses the fact that mandates might simultaneously address both internalities and externalities, even if they address the latter inefficiently. The consequence of missing that fact is to undervalue the potential value, and the potentially high net benefits, of mandates. The third point is that mandates are crude; Choice

Engines might be a preferable alternative, or they might help within the domain authorized by mandates.

Such questions raise a host of empirical issues, to which we lack full answers. Such questions also run into the challenge of heterogeneity. But assuming that many consumers are not paying enough attention to eventual savings in terms of money and time, a suitably designed fuel economy mandate might well be justified, because it would produce an outcome akin to what would be produced by consumers who are not present-biased.[26] Energy efficiency requirements might be justified in similar terms, and indeed, the argument on their behalf might be stronger.[27] If the benefits of mandates greatly exceed their costs, and if there is no significant consumer welfare loss—in the form, for example, of reductions in safety, performance, or aesthetics—then the mandates would seem to serve to correct a behavioral market failure. And indeed, the U.S. government has so argued. Notice the italicized portions of the excerpt here, which appear, in one or another form, in multiple official documents, starting in 2010.

> The central conundrum has been referred to as the *energy paradox* in this setting (and in several others). In short, the problem is that consumers appear not to purchase products that are in their economic self-interest. There are strong theoretical reasons why this might be so:
>
> - *Consumers might be myopic and hence undervalue the long-term.*
> - Consumers might lack information or *a full appreciation of information even when it is presented.*
> - *Consumers might be especially averse to the short-term losses associated with the higher prices of energy-efficient products relative to the uncertain future fuel savings, even if the expected present value of those fuel savings exceeds the cost* (the behavioral phenomenon of *loss aversion*).
> - Even if consumers have relevant knowledge, *the benefits of energy-efficient vehicles might not be sufficiently salient to them at the time of purchase*, and the lack of salience might lead consumers to neglect an attribute that it would be in their economic interest to consider.
> - In the case of vehicle fuel efficiency, and perhaps as a result of one or more of the foregoing factors, consumers may have relatively few choices to purchase vehicles with greater fuel economy once other characteristics, such as vehicle class, are chosen.[28]

As the EPA described the puzzle in 2022:

> A significant question in analyzing consumer impacts from vehicle GHG standards has been why there have appeared to be existing technologies that, if

> adopted, would reduce fuel consumption enough to pay for themselves in short periods, but which were not widely adopted. If the benefits to vehicle buyers outweigh the costs to those buyers of the new technologies, conventional economic principles suggest that automakers would provide them, and people would buy them. Yet engineering analyses have identified a number of technologies whose costs are quickly covered by their fuel savings, such as downsized-turbocharged engines, gasoline direct injection, and improved aerodynamics, that were not widely adopted before the issuance of standards, but which were adopted rapidly afterwards. Why did markets fail, on their own, to adopt these technologies?[29]

Also in 2022, the EPA offered its own answer, pointing

> to consumer behavior, such as putting little emphasis on future fuel savings compared to up-front costs (a form of "myopic loss aversion"), not having a full understanding of potential cost savings, or not prioritizing fuel consumption in the complex process of selecting a vehicle. Explanations of these kinds tend to draw on the conceptual and empirical literature in behavioral economics, which emphasizes the importance of limited attention, the relevance of salience, "present bias" or myopia, and loss aversion. (Some of these are described as contributing to "behavioral market failures.")[30]

At the same time, the EPA was tentative about the relevant research findings, stating that "evidence on technology costs, fuel savings, and the absence of hidden costs suggest that there are market failures in the provision of fuel-saving technologies," while noting that "we cannot demonstrate at this time which specific failures operate in this market."[31]

Of course, we should be cautious before accepting a behavioral argument on behalf of mandates or bans. Present bias has to be demonstrated, not simply asserted; important research suggests that consumers do pay a lot of attention to the benefits of fuel-efficient vehicles.[32] Some of that research finds that with changes in gas prices, consumers adjust their vehicle purchasing decisions, strongly suggesting that in choosing among vehicles, consumers *are* highly attentive to fuel economy.[33] Other research points in the same direction. It finds that when aggressive steps are taken to inform consumers of fuel economy, they do not choose different vehicles.[34]

On the other hand, substantial evidence cuts the other way. Puzzlingly, many consumers do not buy hybrid vehicles even in circumstances in which it would seem rational for them to do so.[35] According to the leading study, a significant number of consumers choose standard vehicles even when it would be in their economic interest to choose a hybrid vehicle, and even when it is difficult to identify some other feature of the standard vehicle that would justify their choosing it.[36]

It is also possible to think that even if consumers are responsive to changes in gasoline prices, they are still myopic with respect to choices of vehicles that have technological advances. Graham et al. put it crisply:

> Consumers are more familiar with changes in fuel price than with changes in technology, since consumers experience fuel prices each time they refill their tank. Vehicle purchases are much less common in the consumer's experience, especially purchases that entail major changes to propulsion systems. Many consumers—excluding the limited pool of adventuresome "early adopters"—may be reticent to purchase vehicles at a premium price that are equipped with unfamiliar engines, transmissions, materials, or entirely new propulsion systems (e.g., hybrids or plug-in electric vehicles), even when such vehicles have attractive EPA fuel-economy ratings.[37]

More broadly, the government's numbers under various presidents, finding no significant consumer welfare loss from fuel economy standards, are consistent with the suggestion that consumers are suffering from some kind of behavioral bias. If consumers were not present biased, we should expect to see some kind of welfare loss, in the form, for example, of vehicles that lacked attributes that consumers preferred.

At the same time, the government's numbers, projecting costs and benefits, may or may not be right. Engineering estimates might overlook some losses that consumers will actually experience along some dimension that they failed to measure. No one doubts that consumers have highly diverse preferences with respect to vehicles, and even though they are not mere defaults, fuel economy standards should be designed to preserve a wide space for freedom of choice. Appropriate standards ensure that such space is maintained. Fuel economy standards do retain considerable space for freedom of choice, and economic incentives have inherent advantages on this count. Both can coexist with Choice Engines, a point to which I will return.

The real question, of course, is the magnitude of net benefits from the different possible approaches. If the consumer savings are taken to be very large, then fuel economy standards are likely to have correspondingly large net benefits. To give a very rough, intuitive sense of how to think about the comparative question, let us suppose that the U.S. government imposed an optimal carbon tax. Simply for purposes of analysis, suppose that it is fifty dollars per ton, understood to capture the social cost of carbon. Suppose that in relevant sectors, including transportation, a certain number of emitters decide to reduce their emissions, on the ground that the cost of reducing them is, on average, $Y, which is lower than fifty dollars. The net benefit of the carbon tax would be fifty dollars minus Y, multiplied by the tons of carbon emissions that are eliminated. It is imaginable that the resulting figure would be very high. But it is not necessarily higher than the net benefits of well-designed fuel economy standards. If consumer savings are real and high, then fuel economy standards might have much higher net benefits than carbon taxes.

## CLIMATE REGULATION AND AI

With the various qualifications, the argument for fuel economy standards, made by reference to present bias and to internalities in general, is at least plausible. In this context, nudges (in the form of an improved fuel economy label) and mandates (in the form of standards) might march hand in hand. It is true that if the goal is only to reduce externalities, a carbon tax is far better than a regulatory mandate. It is also true that in theory, the best approach to internalities should be appropriate disclosure, designed to promote salience and to overcome limited attention. It is also true that a government might respond to present bias with an internality-correcting tax, not with a regulatory mandate. But with an understanding of present bias, a regulatory approach, promoting consumer welfare as well as reducing externalities, might turn out to have higher net benefits than the standard economic remedy of corrective taxes and disclosure.

Everything turns on what the evidence shows, and on the particular numbers. But in principle, regulation of other features of motor vehicles could be also be justified in behavioral terms; present-biased or inattentive consumers might pay too little attention to certain safety at the time of purchase, and some such equipment might fall in the category of experience goods. Credit markets can be analyzed similarly. The broadest point is that while a presumption in favor of freedom of choice makes a great deal of sense, it is only a presumption. If our lodestar is human welfare, it might be overcome, especially when it can be shown that present bias is rampant and that internalities are large.

These have been points about mass regulation. But return to the findings with which I began. Some consumers would benefit from buying electric cars; some would not. Some consumers have a particular taste for electric cars; some do not. Some consumers want large cars; some do not. In any case, cars have a large number of characteristics, and Choice Engines should help people to identify cars with their preferred mix. They should help people to reduce internalities (though stronger medicine might be a good idea[38]). If people want to reduce externalities, Choice Engines could help them to do that as well. Within the limits of mass regulation, private sellers might use Choice Engines. In their ideal form, they would significantly increase social welfare.

## DANGERS AND RISKS

To be sure, there are dangers and risks. Consider three points:

1. Those who design Choice Engines, or anything like them, might be self-interested or malevolent. Rather than correcting an absence of information or behavioral biases, they might *exploit* them. Algorithms and AI threaten to do exactly that, in a way that signals the presence of manipulation.[39] Indeed, AI could turn out to be highly manipulative, thus harming consumers.[40]
2. Choice Engines might turn out to be coarse; they might replicate some of the problems of mass interventions. They may or may not be highly personalized. If they use a few simple cues, such as age and income, they might not have the expected or hoped-for welfare benefits. Algorithms or AI might turn out to be insufficiently informed about the tastes and values of particular choosers.[41]
3. Whether paternalistic or not, AI might turn out to suffer from its own behavioral biases. There is evidence that LLMs show some of the biases that human beings do.[42] It is possible that AI will show biases that human beings show that have not even been named yet. It is also possible that AI will show biases of its own.

For these reasons, the same kinds of guardrails that have been suggested for retirement plans might be applied to Choice Engines of multiple kinds, including those involving motor vehicles, appliances, and energy sources, and thus relating to climate change.[43] Restrictions on the equivalent of *dominated options*, for example, might be imposed by law, so long as it is clear what is dominated.[44] Restrictions on shrouded attributes, including hidden fees, might be similarly justified. Choice Engines, powered by AI, have considerable potential to improve consumer welfare and to dramatically reduce externalities, including those from greenhouse gases. But without regulation, we have reason to question whether they will always or generally do that.[45] Those who design Choice Engines may or may not count as fiduciaries, but at a minimum, it makes sense to scrutinize all forms of choice architecture for deception and manipulation, broadly understood.

# EPILOGUE: THEORY AND PRACTICE

My central conclusions are straightforward:

1. The nations of the world should use the global figure for the social cost of carbon, not the domestic figure. Moral cosmopolitanism, in the constrained form defended here, is sufficient to justify that view. If one nation harms another, it should take that harm into account in deciding what to do.
2. The argument from reciprocity appeals not to morality but to domestic self-interest; it emphasizes that if every nation used the domestic figure, all nations would lose. That argument is convincing. There are no guarantees, of course, that if the United States uses the global number, or if Sweden or Denmark does, other nations will follow suit. But a norm in favor of use of the global number is exceedingly important, and a nation that chooses to use that number can contribute to the creation of the necessary norm. It is worth noting that the argument from moral cosmopolitanism and the argument from reciprocity tend to converge.[1]
3. On welfarist grounds, a degree of redistribution from wealthy nations to poor nations is an excellent idea. A central reason is that a given amount of money is worth a lot more to a poor person than to a wealthy one. If a poor person receives $10,000, it can make a massive difference; if a rich person receives $10,000, it might make no difference at all. It follows that when such redistribution occurs, poor nations are likely to gain more in terms of welfare than wealthy nations are likely to lose.

Apart from that point, there is a good argument that a degree of redistribution is justified from the standpoint of distributive justice.

4. With respect to climate change, these points are complemented by a point about corrective justice. By emitting greenhouse gases, wealthy nations have imposed risks on poorer ones. It is true that in the context of climate change, there is an assortment of problems with both distributive justice and corrective justice arguments. Matters here are intriguingly more complicated than they seem. Even so, those arguments are more right than wrong. Rough justice is still justice.
5. People who are alive now do not deserve greater attention and concern than people who will be born twenty years hence, or forty years hence, or a hundred years hence. Nations should follow a principle of intergenerational neutrality. It does not follow, however, that the discount rate should be zero. Money can be invested and made to grow. If our ancestors used a zero discount rate, we would be a lot worse off than we are now. The right discount rate is a technical matter. But the broader point is clear. It should be based on expected interest rates over the relevant time period. As of this writing, a discount rate in the general vicinity of 2 percent makes sense. If that seems a bit tedious, note well: this point cautions against a discount rate of 5 percent or 7 percent, which some people have advocated, and also against a discount rate near or below 1 percent, which other people have advocated. The stakes are very high.
6. One of the principal goals of climate change policy is to save lives. When nations reduce greenhouse gas emissions or engage in adaptation, a primary goal is to reduce mortality risks. With respect to regulation, valuation of such risks should be rooted in evidence about what informed people, free from behavioral biases, would pay to avoid such risks or would demand to face such risks. In the United States, a VSL of $12 million makes sense (as of the current writing). If China or India used a VSL of $12 million for regulatory purposes, it would be disserving its people.
7. At the same time, subsidies are different from regulation. If the United States, Sweden, Denmark, or France uses a VSL in the vicinity of $12 million for purposes of establishing the social cost of carbon, poor nations would greatly benefit. It is not clear, however, that use of a high VSL is the best way to subsidize poor nations at risk of climate change.

8. Wildfire, drought, extreme heat, and flooding are very much with us, and bound to get worse. Adaptation is receiving increasing attention, and rightly so. The good news is that numerous strategies are available to decrease the relevant risks. Some of those strategies involve the provision of information; some of them involve technical assistance and funding. Here as well, rich nations owe a great deal to poor ones.
9. The problem of climate change is driven in large part by the decisions of consumers. Some of those decisions are terrible for consumers themselves; they cost money and time. They impose large *internalities*, understood as costs that people impose on their future selves. If you buy a product that does not cost a great deal to buy but costs a great deal to operate, then you are imposing internalities on yourself. If you drink alcohol a lot and enjoy drinking a great deal, you will impose internalities on yourself if you end up less healthy. Many consumer decisions relating to everyday products, and not-so-everyday products, involve internalities.
10. Some consumer decisions produce *externalities* in the form of greenhouse gas emissions, where externalities are understood as the costs that people impose on others. Choice Engines, designed with close reference to the climate change problem, can help reduce both internalities and externalities. I have pointed to the importance of guardrails that reduce the risk of self-serving or manipulative Choice Engines.

My focus has largely been on theoretical questions and on matters of right and wrong. But everyone knows that there are limits to how much wealthy nations are willing to do, both in scaling back their emissions and in providing assistance for adaptation. The United States does not want to give a large percentage of its GDP for climate change adaptation, nor do China, the United Arab Emirates, France, Germany, and Canada. Insistence on what is right and what is wrong might derail agreements that are in the interest, above all, of the very nations that are most vulnerable to climate change. That is potentially tragic. International agreements and grants of foreign aid are a product of an unruly mixture of national self-interest and morality.

Still, no one should underrate the importance of ethical judgments, which provide the background against which public officials and negotiators do their work. Those judgments are sometimes a cloud, but they can also be a shining star.

# ACKNOWLEDGMENTS

I have many people to thank. Catherine Woods, my editor, oriented and reoriented the book in multiple ways; in short, she made it much better. David Olin and Victoria Yu were tremendous research assistants. I have discussed these issues for about fifteen years with my wife, Samantha Power, and she much altered my thinking about distributive justice in particular. Still, she should be held responsible for nothing said here, and nothing said here should be taken to reflect her views.

Michael Greenstone has been an inspiration, and a true friend, since 2009, when we worked together in the White House. Eric Posner and David Weisbach, amazing authors and coauthors, have taught me a great deal (and cowrote previous versions of many of the pages here; more on that below). Over the years, discussions with Jon Elster, Robert Hahn, Stephen Holmes, Martha Nussbaum, Richard Posner, Lucia Reisch, Edna Ullmann-Margalit, and Adrian Vermeule have provided terrific help. Three anonymous reviewers improved the book in many ways.

I am honored to have worked in both the Obama administration (in the Office of Information and Regulatory Affairs) and the Biden administration (in the Department of Homeland Security) on climate-related issues. Nothing here reflects an official position in any way (and nothing here was written on government time!). Still, I am grateful to many government colleagues with whom I have discussed some of these issues over the years. I single out seven of my heroes: Carol Browner, Jason Furman, Greenstone, Power, David Hayes, Ali Zaidi, Candace Vahlsing, and Gina McCarthy.

The chapters in this book have various forerunners, in the form of academic articles that have been significantly revised but that provided the foundation for some of the central arguments. Chapter 1 draws on Cass R. Sunstein, *Climate Change Cosmopolitanism*, 39 Yale J. Reg. 1012 (2022). Chapter 2 draws on Eric A. Posner and Cass R. Sunstein, *Climate Change Justice*, 96 Geo. L.J. 1565 (2008). Chapter 3 draws on David Weisbach and Cass R. Sunstein, *Climate Change and Discounting the Future: A Guide for the Perplexed*, 27 Yale L. & Po'y Rev. 433 (2009). Chapter 4 draws on Cass R. Sunstein, *Inequality and the Value of a Statistical Life*, 14 J. Benefit-Cost Analysis 1 (2023). Chapter 5 draws on Cass R. Sunstein, *Foreword: The Imperative of Adapting to Climate Change*, 39 Yale J. Reg. 469 (2022). I am grateful to the relevant journals for permission to revise and adapt the preceding material here.

I am especially grateful to Posner and Weisbach for their collaboration, from which I learned so much. In the chapters that we originally cowrote, much of the prose is theirs. But on some matters, I have ended up reaching very different conclusions from those that the three of us embraced fifteen or so years ago. I do so with fear and trembling.

Special thanks finally to Club Car Cafe in Concord, Massachusetts, where some of this book was written, and to Bianca Giacoppo, who runs the amazing show there.

# APPENDIX: THE PARIS AGREEMENT

*The Parties to this Agreement,*

*Being* Parties to the United Nations Framework Convention on Climate Change, hereinafter referred to as "the Convention,"

*Pursuant* to the Durban Platform for Enhanced Action established by decision 1/CP.17 of the Conference of the Parties to the Convention at its seventeenth session,

*In pursuit* of the objective of the Convention, and being guided by its principles, including the principle of equity and common but differentiated responsibilities and respective capabilities, in the light of different national circumstances,

*Recognizing* the need for an effective and progressive response to the urgent threat of climate change on the basis of the best available scientific knowledge,

*Also recognizing* the specific needs and special circumstances of developing country Parties, especially those that are particularly vulnerable to the adverse effects of climate change, as provided for in the Convention,

*Taking full account* of the specific needs and special situations of the least developed countries with regard to funding and transfer of technology,

*Recognizing* that Parties may be affected not only by climate change, but also by the impacts of the measures taken in response to it,

*Emphasizing* the intrinsic relationship that climate change actions, responses and impacts have with equitable access to sustainable development and eradication of poverty,

*Recognizing* the fundamental priority of safeguarding food security and ending hunger, and the particular vulnerabilities of food production systems to the adverse impacts of climate change,

*Taking into account* the imperatives of a just transition of the workforce and the creation of decent work and quality jobs in accordance with nationally defined development priorities,

*Acknowledging* that climate change is a common concern of humankind, Parties should, when taking action to address climate change, respect, promote and consider their respective obligations on human rights, the right to health, the rights of indigenous peoples, local communities, migrants, children, persons with disabilities and people in vulnerable situations and the right to development, as well as gender equality, empowerment of women and intergenerational equity,

*Recognizing* the importance of the conservation and enhancement, as appropriate, of sinks and reservoirs of the greenhouse gases referred to in the Convention,

*Noting* the importance of ensuring the integrity of all ecosystems, including oceans, and the protection of biodiversity, recognized by some cultures as Mother Earth, and noting the importance for some of the concept of "climate justice," when taking action to address climate change,

*Affirming* the importance of education, training, public awareness, public participation, public access to information and cooperation at all levels on the matters addressed in this Agreement,

*Recognizing* the importance of the engagements of all levels of government and various actors, in accordance with respective national legislations of Parties, in addressing climate change,

*Also recognizing* that sustainable lifestyles and sustainable patterns of consumption and production, with developed country Parties taking the lead, play an important role in addressing climate change,

Have agreed as follows:

## ARTICLE 1

For the purpose of this Agreement, the definitions contained in Article 1 of the Convention shall apply. In addition:

(a) "Convention" means the United Nations Framework Convention on Climate Change, adopted in New York on 9 May 1992;
(b) "Conference of the Parties" means the Conference of the Parties to the Convention;
(c) "Party" means a Party to this Agreement.

## ARTICLE 2

1. This Agreement, in enhancing the implementation of the Convention, including its objective, aims to strengthen the global response to the threat of climate change, in the context of sustainable development and efforts to eradicate poverty, including by:
   (a) Holding the increase in the global average temperature to well below 2 °C above pre-industrial levels and pursuing efforts to limit the temperature increase to 1.5 °C above pre-industrial levels, recognizing that this would significantly reduce the risks and impacts of climate change;
   (b) Increasing the ability to adapt to the adverse impacts of climate change and foster climate resilience and low greenhouse gas emissions development, in a manner that does not threaten food production; and
   (c) Making finance flows consistent with a pathway towards low greenhouse gas emissions and climate-resilient development.
2. This Agreement will be implemented to reflect equity and the principle of common but differentiated responsibilities and respective capabilities, in the light of different national circumstances.

## ARTICLE 3

As nationally determined contributions to the global response to climate change, all Parties are to undertake and communicate ambitious efforts as defined in Articles 4, 7, 9, 10, 11 and 13 with the view to achieving the purpose of this Agreement as set out in Article 2. The efforts of all Parties will represent a progression over time, while recognizing the need to support developing country Parties for the effective implementation of this Agreement.

## ARTICLE 4

1. In order to achieve the long-term temperature goal set out in Article 2, Parties aim to reach global peaking of greenhouse gas emissions as soon as possible, recognizing that peaking will take longer for developing country Parties, and to undertake rapid reductions thereafter in accordance with best available science, so as to achieve a balance between anthropogenic emissions by sources and removals by sinks of greenhouse gases in the second half of this century, on the basis of equity, and in the context of sustainable development and efforts to eradicate poverty.
2. Each Party shall prepare, communicate and maintain successive nationally determined contributions that it intends to achieve. Parties shall pursue domestic mitigation measures, with the aim of achieving the objectives of such contributions.
3. Each Party's successive nationally determined contribution will represent a progression beyond the Party's then current nationally determined contribution and reflect its highest possible ambition, reflecting its common but differentiated responsibilities and respective capabilities, in the light of different national circumstances.
4. Developed country Parties should continue taking the lead by undertaking economy-wide absolute emission reduction targets. Developing country Parties should continue enhancing their mitigation efforts, and are encouraged to move over time towards economy-wide emission reduction or limitation targets in the light of different national circumstances.
5. *Support shall be provided to developing country Parties for the implementation of this Article, in accordance with Articles 9, 10 and 11,* recognizing that enhanced support for developing country Parties will allow for higher ambition in their actions.
6. The least developed countries and small island developing States may prepare and communicate strategies, plans and actions for low greenhouse gas emissions development reflecting their special circumstances.

7. Mitigation co-benefits resulting from Parties' adaptation actions and/or economic diversification plans can contribute to mitigation outcomes under this Article.
8. In communicating their nationally determined contributions, all Parties shall provide the information necessary for clarity, transparency and understanding in accordance with decision 1/CP.21 and any relevant decisions of the Conference of the Parties serving as the meeting of the Parties to this Agreement.
9. Each Party shall communicate a nationally determined contribution every five years in accordance with decision 1/CP.21 and any relevant decisions of the Conference of the Parties serving as the meeting of the Parties to this Agreement and be informed by the outcomes of the global stocktake referred to in Article 14.
10. The Conference of the Parties serving as the meeting of the Parties to this Agreement shall consider common time frames for nationally determined contributions at its first session.
11. A Party may at any time adjust its existing nationally determined contribution with a view to enhancing its level of ambition, in accordance with guidance adopted by the Conference of the Parties serving as the meeting of the Parties to this Agreement.
12. Nationally determined contributions communicated by Parties shall be recorded in a public registry maintained by the secretariat.
13. Parties shall account for their nationally determined contributions. In accounting for anthropogenic emissions and removals corresponding to their nationally determined contributions, Parties shall promote environmental integrity, transparency, accuracy, completeness, comparability and consistency, and ensure the avoidance of double counting, in accordance with guidance adopted by the Conference of the Parties serving as the meeting of the Parties to this Agreement.
14. In the context of their nationally determined contributions, when recognizing and implementing mitigation actions with respect to anthropogenic emissions and removals, Parties should take into account, as appropriate, existing methods and guidance under the Convention, in the light of the provisions of paragraph 13 of this Article.

15. Parties shall take into consideration in the implementation of this Agreement the concerns of Parties with economies most affected by the impacts of response measures, particularly developing country Parties.
16. Parties, including regional economic integration organizations and their member States, that have reached an agreement to act jointly under paragraph 2 of this Article shall notify the secretariat of the terms of that agreement, including the emission level allocated to each Party within the relevant time period, when they communicate their nationally determined contributions. The secretariat shall in turn inform the Parties and signatories to the Convention of the terms of that agreement.
17. Each party to such an agreement shall be responsible for its emission level as set out in the agreement referred to in paragraph 16 of this Article in accordance with paragraphs 13 and 14 of this Article and Articles 13 and 15.
18. If Parties acting jointly do so in the framework of, and together with, a regional economic integration organization which is itself a Party to this Agreement, each member State of that regional economic integration organization individually, and together with the regional economic integration organization, shall be responsible for its emission level as set out in the agreement communicated under paragraph 16 of this Article in accordance with paragraphs 13 and 14 of this Article and Articles 13 and 15.
19. *All Parties should strive to formulate and communicate long-term low greenhouse gas emission development strategies, mindful of Article 2 taking into account* their common but differentiated responsibilities and respective capabilities, in the light of different national circumstances.

## ARTICLE 5

1. Parties should take action to conserve and enhance, as appropriate, sinks and reservoirs of greenhouse gases as referred to in Article 4, *paragraph* 1(d), of the Convention, including forests.
2. Parties are encouraged to take action to implement and support, including through results-based payments, the existing framework as set out in

related guidance and decisions already agreed under the Convention for: policy approaches and positive incentives for activities relating to reducing emissions from deforestation and forest degradation, and the role of conservation, sustainable management of forests and enhancement of forest carbon stocks in developing countries; and alternative policy approaches, such as joint mitigation and adaptation approaches for the integral and sustainable management of forests, while reaffirming the importance of incentivizing, as appropriate, non-carbon benefits associated with such approaches.

## ARTICLE 6

1. Parties recognize that some Parties choose to pursue voluntary cooperation in the implementation of their nationally determined contributions to allow for higher ambition in their mitigation and adaptation actions and to promote sustainable development and environmental integrity.
2. Parties shall, where engaging on a voluntary basis in cooperative approaches that involve the use of internationally transferred mitigation outcomes towards nationally determined contributions, promote sustainable development and ensure environmental integrity and transparency, including in governance, and shall apply robust accounting to ensure, inter alia, the avoidance of double counting, consistent with guidance adopted by the Conference of the Parties serving as the meeting of the Parties to this Agreement.
3. The use of internationally transferred mitigation outcomes to achieve nationally determined contributions under this Agreement shall be voluntary and authorized by participating Parties.
4. A mechanism to contribute to the mitigation of greenhouse gas emissions and support sustainable development is hereby established under the authority and guidance of the Conference of the Parties serving as the meeting of the Parties to this Agreement for use by Parties on a voluntary basis. It shall be supervised by a body designated by the Conference of the Parties serving as the meeting of the Parties to this Agreement, and shall aim:
   (a) To promote the mitigation of greenhouse gas emissions while fostering sustainable development;

(b) To incentivize and facilitate participation in the mitigation of greenhouse gas emissions by public and private entities authorized by a Party;

(c) To contribute to the reduction of emission levels in the host Party, which will benefit from mitigation activities resulting in emission reductions that can also be used by another Party to fulfil its nationally determined contribution; and

(d) To deliver an overall mitigation in global emissions.

5. Emission reductions resulting from the mechanism referred to in paragraph 4 of this Article shall not be used to demonstrate achievement of the host Party's nationally determined contribution if used by another Party to demonstrate achievement of its nationally determined contribution.

6. The Conference of the Parties serving as the meeting of the Parties to this Agreement shall ensure that a share of the proceeds from activities under the mechanism referred to in paragraph 4 of this Article is used to cover administrative expenses as well as to assist developing country Parties that are particularly vulnerable to the adverse effects of climate change to meet the costs of adaptation.

7. The Conference of the Parties serving as the meeting of the Parties to this Agreement shall adopt rules, modalities and procedures for the mechanism referred to in paragraph 4 of this Article at its first session.

8. Parties recognize the importance of integrated, holistic and balanced non-market approaches being available to Parties to assist in the implementation of their nationally determined contributions, in the context of sustainable development and poverty eradication, in a coordinated and effective manner, including through, inter alia, mitigation, adaptation, finance, technology transfer and capacity-building, as appropriate. These approaches shall aim to:

(a) Promote mitigation and adaptation ambition;

(b) Enhance public and private sector participation in the implementation of nationally determined contributions; and

(c) Enable opportunities for coordination across instruments and relevant institutional arrangements.

9. A framework for non-market approaches to sustainable development is hereby defined to promote the non-market approaches referred to in paragraph 8 of this Article.

## ARTICLE 7

1. Parties hereby establish the global goal on adaptation of enhancing adaptive capacity, strengthening resilience and reducing vulnerability to climate change, with a view to contributing to sustainable development and ensuring an adequate adaptation response in the context of the temperature goal referred to in Article 2.
2. Parties recognize that adaptation is a global challenge faced by all with local, subnational, national, regional and international dimensions, and that it is a key component of and makes a contribution to the long-term global response to climate change to protect people, livelihoods and ecosystems, taking into account the urgent and immediate needs of those developing country Parties that are particularly vulnerable to the adverse effects of climate change.
3. The adaptation efforts of developing country Parties shall be recognized, in accordance with the modalities to be adopted by the Conference of the Parties serving as the meeting of the Parties to this Agreement at its first session.
4. Parties recognize that the current need for adaptation is significant and that greater levels of mitigation can reduce the need for additional adaptation efforts, and that greater adaptation needs can involve greater adaptation costs.
5. Parties acknowledge that adaptation action should follow a country-driven, gender-responsive, participatory and fully transparent approach, taking into consideration vulnerable groups, communities and ecosystems, and should be based on and guided by the best available science and, as appropriate, traditional knowledge, knowledge of indigenous peoples and local knowledge systems, with a view to integrating adaptation into relevant socioeconomic and environmental policies and actions, where appropriate.
6. Parties recognize the importance of support for and international cooperation on adaptation efforts and the importance of taking into account

the needs of developing country Parties, especially those that are particularly vulnerable to the adverse effects of climate change.

7. Parties should strengthen their cooperation on enhancing action on adaptation, taking into account the Cancun Adaptation Framework, including with regard to:
   (a) Sharing information, good practices, experiences and lessons learned, including, as appropriate, as these relate to science, planning, policies and implementation in relation to adaptation actions;
   (b) Strengthening institutional arrangements, including those under the Convention that serve this Agreement, to support the synthesis of relevant information and knowledge, and the provision of technical support and guidance to Parties;
   (c) Strengthening scientific knowledge on climate, including research, systematic observation of the climate system and early warning systems, in a manner that informs climate services and supports decision-making;
   (d) Assisting developing country Parties in identifying effective adaptation practices, adaptation needs, priorities, support provided and received for adaptation actions and efforts, and challenges and gaps, in a manner consistent with encouraging good practices; and
   (e) Improving the effectiveness and durability of adaptation actions.
8. United Nations specialized organizations and agencies are encouraged to support the efforts of Parties to implement the actions referred to in paragraph 7 of this Article, taking into account the provisions of paragraph 5 of this Article.
9. Each Party shall, as appropriate, engage in adaptation planning processes and the implementation of actions, including the development or enhancement of relevant plans, policies and/or contributions, which may include:
   (a) The implementation of adaptation actions, undertakings and/or efforts;
   (b) The process to formulate and implement national adaptation plans;
   (c) The assessment of climate change impacts and vulnerability, with a view to formulating nationally determined prioritized actions, taking into account vulnerable people, places and ecosystems;

(d) Monitoring and evaluating and learning from adaptation plans, policies, programmes and actions; and

(e) Building the resilience of socioeconomic and ecological systems, including through economic diversification and sustainable management of natural resources.

10. Each Party should, as appropriate, submit and update periodically an adaptation communication, which may include its priorities, implementation and support needs, plans and actions, without creating any additional burden for developing country Parties.
11. The adaptation communication referred to in paragraph 10 of this Article shall be, as appropriate, submitted and updated periodically, as a component of or in conjunction with other communications or documents, including a national adaptation plan, a nationally determined contribution as referred to in Article 4, paragraph 2, and/or a national communication.
12. The adaptation communications referred to in paragraph 10 of this Article shall be recorded in a public registry maintained by the secretariat.
13. Continuous and enhanced international support shall be provided to developing country Parties for the implementation of paragraphs 7, 9, 10 and 11 of this Article, in accordance with the provisions of Articles 9, 10 and 11.
14. The global stocktake referred to in Article 14 shall, inter alia:
    (a) Recognize adaptation efforts of developing country Parties;
    (b) Enhance the implementation of adaptation action taking into account the adaptation communication referred to in paragraph 10 of this Article;
    (c) Review the adequacy and effectiveness of adaptation and support provided for adaptation; and
    (d) Review the overall progress made in achieving the global goal on adaptation referred to in paragraph 1 of this Article.

## ARTICLE 8

1. Parties recognize the importance of averting, minimizing and addressing loss and damage associated with the adverse effects of climate change,

including extreme weather events and slow onset events, and the role of sustainable development in reducing the risk of loss and damage.

2. The Warsaw International Mechanism for Loss and Damage associated with Climate Change Impacts shall be subject to the authority and guidance of the Conference of the Parties serving as the meeting of the Parties to this Agreement and may be enhanced and strengthened, as determined by the Conference of the Parties serving as the meeting of the Parties to this Agreement.
3. Parties should enhance understanding, action and support, including through the Warsaw International Mechanism, as appropriate, on a cooperative and facilitative basis with respect to loss and damage associated with the adverse effects of climate change.
4. Accordingly, areas of cooperation and facilitation to enhance understanding, action and support may include:
   (a) Early warning systems;
   (b) Emergency preparedness;
   (c) Slow onset events;
   (d) Events that may involve irreversible and permanent loss and damage;
   (e) Comprehensive risk assessment and management;
   (f) Risk insurance facilities, climate risk pooling and other insurance solutions;
   (g) Non-economic losses; and
   (h) Resilience of communities, livelihoods and ecosystems.
5. The Warsaw International Mechanism shall collaborate with existing bodies and expert groups under the Agreement, as well as relevant organizations and expert bodies outside the Agreement.

## ARTICLE 9

1. Developed country Parties shall provide financial resources to assist developing country Parties with respect to both mitigation and adaptation in continuation of their existing obligations under the Convention.
2. Other Parties are encouraged to provide or continue to provide such support voluntarily.

3. As part of a global effort, developed country Parties should continue to take the lead in mobilizing climate finance from a wide variety of sources, instruments and channels, noting the significant role of public funds, through a variety of actions, including supporting country-driven strategies, and taking into account the needs and priorities of developing country Parties. Such mobilization of climate finance should represent a progression beyond previous efforts.
4. The provision of scaled-up financial resources should aim to achieve a balance between adaptation and mitigation, taking into account country-driven strategies, and the priorities and needs of developing country Parties, especially those that are particularly vulnerable to the adverse effects of climate change and have significant capacity constraints, such as the least developed countries and small island developing States, considering the need for public and grant-based resources for adaptation.
5. Developed country Parties shall biennially communicate indicative quantitative and qualitative information related to paragraphs 1 and 3 of this Article, as applicable, including, as available, projected levels of public financial resources to be provided to developing country Parties. Other Parties providing resources are encouraged to communicate biennially such information on a voluntary basis.
6. The global stocktake referred to in Article 14 shall take into account the relevant information provided by developed country Parties and/or Agreement bodies on efforts related to climate finance.
7. Developed country Parties shall provide transparent and consistent information on support for developing country Parties provided and mobilized through public interventions biennially in accordance with the modalities, procedures and guidelines to be adopted by the Conference of the Parties serving as the meeting of the Parties to this Agreement, at its first session, as stipulated in Article 13, paragraph 13. Other Parties are encouraged to do so.
8. The Financial Mechanism of the Convention, including its operating entities, shall serve as the financial mechanism of this Agreement.
9. The institutions serving this Agreement, including the operating entities of the Financial Mechanism of the Convention, shall aim to ensure efficient access to financial resources through simplified approval

procedures and enhanced readiness support for developing country Parties, in particular for the least developed countries and small island developing States, in the context of their national climate strategies and plans.

## ARTICLE 10

1. Parties share a long-term vision on the importance of fully realizing technology development and transfer in order to improve resilience to climate change and to reduce greenhouse gas emissions.
2. Parties, noting the importance of technology for the implementation of mitigation and adaptation actions under this Agreement and recognizing existing technology deployment and dissemination efforts, shall strengthen cooperative action on technology development and transfer.
3. The Technology Mechanism established under the Convention shall serve this Agreement.
4. A technology framework is hereby established to provide overarching guidance to the work of the Technology Mechanism in promoting and facilitating enhanced action on technology development and transfer in order to support the implementation of this Agreement, in pursuit of the long-term vision referred to in paragraph 1 of this Article.
5. Accelerating, encouraging and enabling innovation is critical for an effective, long-term global response to climate change and promoting economic growth and sustainable development. Such effort shall be, as appropriate, supported, including by the Technology Mechanism and, through financial means, by the Financial Mechanism of the Convention, for collaborative approaches to research and development, and facilitating access to technology, in particular for early stages of the technology cycle, to developing country Parties.
6. Support, including financial support, shall be provided to developing country Parties for the implementation of this Article, including for strengthening cooperative action on technology development and transfer at different stages of the technology cycle, with a view to achieving a balance between support for mitigation and adaptation. The global stocktake referred to in Article 14 shall take into account available

information on efforts related to support on technology development and transfer for developing country Parties.

## ARTICLE 11

1. Capacity-building under this Agreement should enhance the capacity and ability of developing country Parties, in particular countries with the least capacity, such as the least developed countries, and those that are particularly vulnerable to the adverse effects of climate change, such as small island developing States, to take effective climate change action, including, inter alia, to implement adaptation and mitigation actions, and should facilitate technology development, dissemination and deployment, access to climate finance, relevant aspects of education, training and public awareness, and the transparent, timely and accurate communication of information.
2. Capacity-building should be country-driven, based on and responsive to national needs, and foster country ownership of Parties, in particular, for developing country Parties, including at the national, subnational and local levels. Capacity-building should be guided by lessons learned, including those from capacity-building activities under the Convention, and should be an effective, iterative process that is participatory, cross-cutting and gender-responsive.
3. All Parties should cooperate to enhance the capacity of developing country Parties to implement this Agreement. Developed country Parties should enhance support for capacity-building actions in developing country Parties.
4. All Parties enhancing the capacity of developing country Parties to implement this Agreement, including through regional, bilateral and multilateral approaches, shall regularly communicate on these actions or measures on capacity-building. Developing country Parties should regularly communicate progress made on implementing capacity-building plans, policies, actions or measures to implement this Agreement.
5. Capacity-building activities shall be enhanced through appropriate institutional arrangements to support the implementation of this Agreement, including the appropriate institutional arrangements established under the Convention that serve this Agreement. The Conference of the

Parties serving as the meeting of the Parties to this Agreement shall, at its first session, consider and adopt a decision on the initial institutional arrangements for capacity-building.

ARTICLE 12

Parties shall cooperate in taking measures, as appropriate, to enhance climate change education, training, public awareness, public participation and public access to information, recognizing the importance of these steps with respect to enhancing actions under this Agreement.

ARTICLE 13

1. In order to build mutual trust and confidence and to promote effective implementation, an enhanced transparency framework for action and support, with built-in flexibility which takes into account Parties' different capacities and builds upon collective experience is hereby established.
2. The transparency framework shall provide flexibility in the implementation of the provisions of this Article to those developing country Parties that need it in the light of their capacities. The modalities, procedures and guidelines referred to in paragraph 13 of this Article shall reflect such flexibility.
3. The transparency framework shall build on and enhance the transparency arrangements under the Convention, recognizing the special circumstances of the least developed countries and small island developing States, and be implemented in a facilitative, non-intrusive, non-punitive manner, respectful of national sovereignty, and avoid placing undue burden on Parties.
4. The transparency arrangements under the Convention, including national communications, biennial reports and biennial update reports, international assessment and review and international consultation and analysis, shall form part of the experience drawn upon for the development of the modalities, procedures and guidelines under paragraph 13 of this Article.

5. The purpose of the framework for transparency of action is to provide a clear understanding of climate change action in the light of the objective of the Convention as set out in its Article 2, including clarity and tracking of progress towards achieving Parties' individual nationally determined contributions under Article 4, and Parties' adaptation actions under Article 7, including good practices, priorities, needs and gaps, to inform the global stocktake under Article 14.
6. The purpose of the framework for transparency of support is to provide clarity on support provided and received by relevant individual Parties in the context of climate change actions under Articles 4, 7, 9, 10 and 11, and, to the extent possible, to provide a full overview of aggregate financial support provided, to inform the global stocktake under Article 14.
7. Each Party shall regularly provide the following information:
   (a) A national inventory report of anthropogenic emissions by sources and removals by sinks of greenhouse gases, prepared using good practice methodologies accepted by the Intergovernmental Panel on Climate Change and agreed upon by the Conference of the Parties serving as the meeting of the Parties to this Agreement; and
   (b) Information necessary to track progress made in implementing and achieving its nationally determined contribution under Article 4.
8. Each Party should also provide information related to climate change impacts and adaptation under Article 7, as appropriate.
9. Developed country Parties shall, and other Parties that provide support should, provide information on financial, technology transfer and capacity-building support provided to developing country Parties under Articles 9, 10 and 11.
10. Developing country Parties should provide information on financial, technology transfer and capacity-building support needed and received under Articles 9, 10 and 11.
11. Information submitted by each Party under paragraphs 7 and 9 of this Article shall undergo a technical expert review, in accordance with decision 1/CP.21. For those developing country Parties that need it in the light of their capacities, the review process shall include assistance in identifying capacity-building needs. In addition, each Party shall

participate in a facilitative, multilateral consideration of progress with respect to efforts under Article 9, and its respective implementation and achievement of its nationally determined contribution.

12. The technical expert review under this paragraph shall consist of a consideration of the Party's support provided, as relevant, and its implementation and achievement of its nationally determined contribution. The review shall also identify areas of improvement for the Party, and include a review of the consistency of the information with the modalities, procedures and guidelines referred to in paragraph 13 of this Article, taking into account the flexibility accorded to the Party under paragraph 2 of this Article. The review shall pay particular attention to the respective national capabilities and circumstances of developing country Parties.
13. The Conference of the Parties serving as the meeting of the Parties to this Agreement shall, at its first session, building on experience from the arrangements related to transparency under the Convention, and elaborating on the provisions in this Article, adopt common modalities, procedures and guidelines, as appropriate, for the transparency of action and support.
14. Support shall be provided to developing countries for the implementation of this Article.
15. Support shall also be provided for the building of transparency-related capacity of developing country Parties on a continuous basis.

## ARTICLE 14

1. The Conference of the Parties serving as the meeting of the Parties to this Agreement shall periodically take stock of the implementation of this Agreement to assess the collective progress towards achieving the purpose of this Agreement and its long-term goals (referred to as the "global stocktake"). It shall do so in a comprehensive and facilitative manner, considering mitigation, adaptation and the means of implementation and support, and in the light of equity and the best available science.
2. The Conference of the Parties serving as the meeting of the Parties to this Agreement shall undertake its first global stocktake in 2023 and every

five years thereafter unless otherwise decided by the Conference of the Parties serving as the meeting of the Parties to this Agreement.

3. The outcome of the global stocktake shall inform Parties in updating and enhancing, in a nationally determined manner, their actions and support in accordance with the relevant provisions of this Agreement, as well as in enhancing international cooperation for climate action.

## ARTICLE 15

1. A mechanism to facilitate implementation of and promote compliance with the provisions of this Agreement is hereby established.
2. The mechanism referred to in paragraph 1 of this Article shall consist of a committee that shall be expert-based and facilitative in nature and function in a manner that is transparent, non-adversarial and non-punitive. The committee shall pay particular attention to the respective national capabilities and circumstances of Parties.
3. The committee shall operate under the modalities and procedures adopted by the Conference of the Parties serving as the meeting of the Parties to this Agreement at its first session and report annually to the Conference of the Parties serving as the meeting of the Parties to this Agreement.

## ARTICLE 16

1. The Conference of the Parties, the supreme body of the Convention, shall serve as the meeting of the Parties to this Agreement.
2. Parties to the Convention that are not Parties to this Agreement may participate as observers in the proceedings of any session of the Conference of the Parties serving as the meeting of the Parties to this Agreement. When the Conference of the Parties serves as the meeting of the Parties to this Agreement, decisions under this Agreement shall be taken only by those that are Parties to this Agreement.
3. When the Conference of the Parties serves as the meeting of the Parties to this Agreement, any member of the Bureau of the Conference of the Parties representing a Party to the Convention but, at that time, not a

Party to this Agreement, shall be replaced by an additional member to be elected by and from amongst the Parties to this Agreement.

4. The Conference of the Parties serving as the meeting of the Parties to this Agreement shall keep under regular review the implementation of this Agreement and shall make, within its mandate, the decisions necessary to promote its effective implementation. It shall perform the functions assigned to it by this Agreement and shall:
   (a) Establish such subsidiary bodies as deemed necessary for the implementation of this Agreement; and
   (b) Exercise such other functions as may be required for the implementation of this Agreement.
5. The rules of procedure of the Conference of the Parties and the financial procedures applied under the Convention shall be applied *mutatis mutandis* under this Agreement, except as may be otherwise decided by consensus by the Conference of the Parties serving as the meeting of the Parties to this Agreement.
6. The first session of the Conference of the Parties serving as the meeting of the Parties to this Agreement shall be convened by the secretariat in conjunction with the first session of the Conference of the Parties that is scheduled after the date of entry into force of this Agreement. Subsequent ordinary sessions of the Conference of the Parties serving as the meeting of the Parties to this Agreement shall be held in conjunction with ordinary sessions of the Conference of the Parties, unless otherwise decided by the Conference of the Parties serving as the meeting of the Parties to this Agreement.
7. Extraordinary sessions of the Conference of the Parties serving as the meeting of the Parties to this Agreement shall be held at such other times as may be deemed necessary by the Conference of the Parties serving as the meeting of the Parties to this Agreement or at the written request of any Party, provided that, within six months of the request being communicated to the Parties by the secretariat, it is supported by at least one third of the Parties.
8. The United Nations and its specialized agencies and the International Atomic Energy Agency, as well as any State member thereof or observers thereto not party to the Convention, may be represented at sessions of

the Conference of the Parties serving as the meeting of the Parties to this Agreement as observers. Any body or agency, whether national or international, governmental or non-governmental, which is qualified in matters covered by this Agreement and which has informed the secretariat of its wish to be represented at a session of the Conference of the Parties serving as the meeting of the Parties to this Agreement as an observer, may be so admitted unless at least one third of the Parties present object. The admission and participation of observers shall be subject to the rules of procedure referred to in paragraph 5 of this Article.

## ARTICLE 17

1. The secretariat established by Article 8 of the Convention shall serve as the secretariat of this Agreement.
2. Article 8, paragraph 2, of the Convention on the functions of the secretariat, and Article 8, paragraph 3, of the Convention, on the arrangements made for the functioning of the secretariat, shall apply *mutatis mutandis* to this Agreement. The secretariat shall, in addition, exercise the functions assigned to it under this Agreement and by the Conference of the Parties serving as the meeting of the Parties to this Agreement.

## ARTICLE 18

1. The Subsidiary Body for Scientific and Technological Advice and the Subsidiary Body for Implementation established by Articles 9 and 10 of the Convention shall serve, respectively, as the Subsidiary Body for Scientific and Technological Advice and the Subsidiary Body for Implementation of this Agreement. The provisions of the Convention relating to the functioning of these two bodies shall apply *mutatis mutandis* to this Agreement. Sessions of the meetings of the Subsidiary Body for Scientific and Technological Advice and the Subsidiary Body for Implementation of this Agreement shall be held in conjunction with the meetings of, respectively, the Subsidiary Body for Scientific and Technological Advice and the Subsidiary Body for Implementation of the Convention.
2. Parties to the Convention that are not Parties to this Agreement may participate as observers in the proceedings of any session of the subsidiary

bodies. When the subsidiary bodies serve as the subsidiary bodies of this Agreement, decisions under this Agreement shall be taken only by those that are Parties to this Agreement.

3. When the subsidiary bodies established by Articles 9 and 10 of the Convention exercise their functions with regard to matters concerning this Agreement, any member of the bureaux of those subsidiary bodies representing a Party to the Convention but, at that time, not a Party to this Agreement, shall be replaced by an additional member to be elected by and from amongst the Parties to this Agreement.

## ARTICLE 19

1. Subsidiary bodies or other institutional arrangements established by or under the Convention, other than those referred to in this Agreement, shall serve this Agreement upon a decision of the Conference of the Parties serving as the meeting of the Parties to this Agreement. The Conference of the Parties serving as the meeting of the Parties to this Agreement shall specify the functions to be exercised by such subsidiary bodies or arrangements.
2. The Conference of the Parties serving as the meeting of the Parties to this Agreement may provide further guidance to such subsidiary bodies and institutional arrangements.

## ARTICLE 20

1. This Agreement shall be open for signature and subject to ratification, acceptance or approval by States and regional economic integration organizations that are Parties to the Convention. It shall be open for signature at the United Nations Headquarters in New York from 22 April 2016 to 21 April 2017. Thereafter, this Agreement shall be open for accession from the day following the date on which it is closed for signature. Instruments of ratification, acceptance, approval or accession shall be deposited with the Depositary.
2. Any regional economic integration organization that becomes a Party to this Agreement without any of its member States being a Party shall

be bound by all the obligations under this Agreement. In the case of regional economic integration organizations with one or more member States that are Parties to this Agreement, the organization and its member States shall decide on their respective responsibilities for the performance of their obligations under this Agreement. In such cases, the organization and the member States shall not be entitled to exercise rights under this Agreement concurrently.

3. In their instruments of ratification, acceptance, approval or accession, regional economic integration organizations shall declare the extent of their competence with respect to the matters governed by this Agreement. These organizations shall also inform the Depositary, who shall in turn inform the Parties, of any substantial modification in the extent of their competence.

## ARTICLE 21

1. This Agreement shall enter into force on the thirtieth day after the date on which at least 55 Parties to the Convention accounting in total for at least an estimated 55 per cent of the total global greenhouse gas emissions have deposited their instruments of ratification, acceptance, approval or accession.
2. Solely for the limited purpose of paragraph 1 of this Article, "total global greenhouse gas emissions" means the most up-to-date amount communicated on or before the date of adoption of this Agreement by the Parties to the Convention.
3. For each State or regional economic integration organization that ratifies, accepts or approves this Agreement or accedes thereto after the conditions set out in paragraph 1 of this Article for entry into force have been fulfilled, this Agreement shall enter into force on the thirtieth day after the date of deposit by such State or regional economic integration organization of its instrument of ratification, acceptance, approval or accession.
4. For the purposes of paragraph 1 of this Article, any instrument deposited by a regional economic integration organization shall not be counted as additional to those deposited by its member States.

## ARTICLE 22

The provisions of Article 15 of the Convention on the adoption of amendments to the Convention shall apply *mutatis mutandis* to this Agreement.

## ARTICLE 23

1. The provisions of Article 16 of the Convention on the adoption and amendment of annexes to the Convention shall apply *mutatis mutandis* to this Agreement.
2. Annexes to this Agreement shall form an integral part thereof and, unless otherwise expressly provided for, a reference to this Agreement constitutes at the same time a reference to any annexes thereto. Such annexes shall be restricted to lists, forms and any other material of a descriptive nature that is of a scientific, technical, procedural or administrative character.

## ARTICLE 24

The provisions of Article 14 of the Convention on settlement of disputes shall apply *mutatis mutandis* to this Agreement.

## ARTICLE 25

1. Each Party shall have one vote, except as provided for in paragraph 2 of this Article.
2. Regional economic integration organizations, in matters within their competence, shall exercise their right to vote with a number of votes equal to the number of their member States that are Parties to this Agreement. Such an organization shall not exercise its right to vote if any of its member States exercises its right, and vice versa.

## ARTICLE 26

The Secretary-General of the United Nations shall be the Depositary of this Agreement.

## ARTICLE 27

No reservations may be made to this Agreement.

## ARTICLE 28

1. At any time after three years from the date on which this Agreement has entered into force for a Party, that Party may withdraw from this Agreement by giving written notification to the Depositary.
2. Any such withdrawal shall take effect upon expiry of one year from the date of receipt by the Depositary of the notification of withdrawal, or on such later date as may be specified in the notification of withdrawal.
3. Any Party that withdraws from the Convention shall be considered as also having withdrawn from this Agreement.

## ARTICLE 29

The original of this Agreement, of which the Arabic, Chinese, English, French, Russian and Spanish texts are equally authentic, shall be deposited with the Secretary-General of the United Nations.

DONE at Paris this twelfth day of December two thousand and fifteen.

IN WITNESS WHEREOF, the undersigned, being duly authorized to that effect, have signed this Agreement.

# NOTES

## INTRODUCTION

1. Christopher W. Callahan & Justin S. Mankin, *National Attribution of Historical Climate Change Damages*, 172 Climate Change 40 (2022).

2. John Stuart Mill, *Utilitarianism* 24–25 (Parker, Son, & Bourn, 1863) (1861).

3. See Eric Posner & David Weisbach, *Climate Change Justice* (2009).

4. On some of these questions, see Amartya Sen, *Utilitarianism and Welfarism*, 76 J. Phil. 463 (1979).

## CHAPTER 1

1. William Blake, *The Complete Poetry and Prose of William Blake* 641 (David Erdman ed., 1988).

2. Exec. Order No. 12866, 58 Fed. Reg. 51,735 (Oct. 4, 1993).

3. Exec. Order No. 13563, 76 Fed. Reg. 3,821 (Jan. 21, 2011).

4. See Cass R. Sunstein, *The Cost-Benefit Revolution* (2018).

5. Eric Roston, *The Most Important Number You've Never Heard Of*, Bloomberg (Jan. 22, 2021), https://www.bloomberg.com/news/articles/2021-01-22/how-do-you-put-a-price-on-climate-change-michael-greenstone-knows.

6. Tamma Carleton & Michael Greenstone, *Updating the United States Government's Social Cost of Carbon* (Univ. of Chi., Becker Friedman Inst. for Econ., Working Paper No. 2021-04, 2021).

7. See Env. Prot. Agency, *Report on the Social Cost of Greenhouse Gases: Estimates Incorporating Recent Scientific Advances* (2022).

8. Peter Howard and Jason Schwartz, *Think Global: International Reciprocity as Justification for a Global Social Cost of Carbon*, 42 Colum. J. Env't L. 203, 219–220, 270–294 (2017).

9. The literature is vast. See, e.g., National Academy of Sciences, Engineering, and Medicine, *Valuing Climate Damages: Updating Estimation of the Social Cost of Carbon Dioxide* (2017); Katharine Ricke, Laurent Drouet, Ken Caldeira, & Massimo Tavoni, *Country-Level Social Cost of Carbon*, 8 Nature Climate Change 895 (2018); William D. Nordhaus, *Revisiting the Social Cost of Carbon*, 114 Proc. Nat'l Acad. Sci. U.S. 1518 (2017); Carleton & Greenstone, supra note 6; Gernot Wagner, David Anthoff, Maureen Cropper, Simon Dietz, Kenneth T. Gillingham, Ben Groom, J. Paul Kelleher, Frances C. Moore, & James H. Stock, *Comment: Eight Priorities for Calculating the Social Cost of Carbon*, 590 Nature 548 (2021); Nicholas Stern & Joseph E. Stiglitz, *The Social Cost of Carbon, Risk, Distribution, Market Failures: An Alternative Approach* (Nat'l Bureau of Econ. Rsch., Working Paper No. 28472, 2021); Michael Greenstone, Elizabeth Kopitz, & Ann Wolverton, *Estimating the Social Cost of Carbon for Use in U.S. Federal Rulemakings: A Summary and Interpretation* (Nat'l Bureau of Econ. Rsch., Working Paper No. 16913, 2011); Jonathan S. Masur & Eric A. Posner, *Climate Regulation and the Limits of Cost-Benefit Analysis*, 99 Calif. L. Rev. 1557 (2011); Robert S. Pindyck, *The Social Cost of Carbon Revisited*, 94 J. Env't Econ. & Mgmt. 140 (2019); Howard & Schwartz, supra note 8; Ted Gayer & W. Kip Viscusi, *Determining the Proper Scope of Climate Change Policy Benefits in U.S. Regulatory Analyses: Domestic Versus Global Approaches*, 10 Rev. Env't. Econ. & Pol'y 245 (2016).

10. See Carleton & Greenstone, supra note 6, at 10.

11. Id. at 13.

12. See Matthew J. Kotchen, *Which Social Cost of Carbon? A Theoretical Perspective*, 5 J. Ass'n Env't & Res. Econ. 673 (2018); Howard & Schwartz, supra note 8.

13. There is a complex and important question in the background. If the United States values foreign lives, in the sense that it includes premature deaths to foreigners in its benefit calculations, at what level should it value those lives? I take up this question in chapter 4.

14. I use the term *moral cosmopolitanism* to refer very simply to the idea that a nation ought to value foreign lives, not only the lives of its own citizens, or even those within its territorial boundaries. See Arden Rowell & Lesley Wexler, *Valuing Foreign Lives*, 48 Ga. L. Rev. 499 (2014). I do not mean to say anything provocative or controversial about the cosmopolitan tradition, or about what cosmopolitanism ought to be taken to entail. For a valuable discussion, see Martha Nussbaum, *The Cosmopolitan Tradition: A Noble but Flawed Idea* (2019).

15. See generally John Rawls, *A Theory of Justice* (1971), and in particular the treatment of the original position and the veil of ignorance. I do not mean to suggest that Rawls's discussion is meant to answer the question posed in the text.

16. Cf. Randall A. Kramer & D. Evan Mercer, *Valuing a Global Environmental Good: U.S. Residents' Willingness to Pay to Protect Tropical Rain Forests*, 73 Land Econ. 196 (1997), exploring the willingness of Americans to pay for tropical rainforest protection.

17. Off. of Mgmt. & Budget, Exec. Off. of the President, Circular A-4, To the Heads of Executive Agencies and Establishments: Regulatory Analysis 15 (2003). In 2023, Circular A-4 was revised. See Cass R. Sunstein, *The Economic Constitution of the United States*, J. Econ. Persp. (forthcoming 2024); Off. of Mgmt. & Budget, Exec. Off. of the President, Circular A-4 (2023), available at https://www.whitehouse.gov/wp-content/uploads/2023/11/CircularA-4.pdf.

18. See Rowell & Wexler, supra note 14, at 522–532.

19. Id. at 526.

20. Id. at 531.

21. On some of those, see Howard & Schwartz, supra note 8, at 211–219; Masur & Posner, supra note 9; Michael Greenstone, Elizabeth Kopits, & Ann Wolverton, *Developing a Social Cost of Carbon for US Regulatory Analysis: A Methodology and Interpretation*, 7 Rev. Env't Econ. & Pol'y 23, 23 (2011).

22. See California v. Bernhard, 472 F. Supp. 3d 573 (N.D. Cal. 2020).

23. Interagency Working Grp. on Soc. Cost of Carbon, U.S. Gov't, *Technical Support Document: Social Cost of Carbon for Regulatory Impact Analysis under Executive Order 12866* (2010), hereinafter Interagency Working Grp. (2010).

24. See Interagency Working Grp. (2010), supra note 23, at 10–11; Interagency Working Grp. on Soc. Cost of Carbon, U.S. Gov't, *Technical Update of the Social Cost of Carbon for Regulatory Impact Analysis under Executive Order 12866*, 17 (2016), hereinafter Interagency Working Grp. (2016).

25. Interagency Working Grp. (2010), supra note 23, at 10.

26. Id. at 11.

27. Interagency Working Grp. (2016), supra note 24, at 17.

28. Id.

29. Id.

30. Id.

31. Id.

32. Id. at 15.

33. See Kotchen, supra note 12.

34. Interagency Working Grp. on. Soc. Cost of Carbon, U.S. Gov't, *Technical Support Document: Social Sot of Carbon, Methane, and Nitrous Oxide, Interim Estimates Under Executive Order 13990* 16 (2021), hereinafter Interagency Working Grp. (2021).

35. Id.

36. Exec. Order No. 13783, 82 Fed. Reg. 16,093 (Mar. 28, 2017).

37. Id. § 4(c).

38. Exec. Order No. 13990, 86 Fed. Reg. 7,037 (Jan. 20, 2021).

39. Interagency Working Grp. (2021), supra note 34.

40. Id. at 15–16.

41. Env. Prot. Agency, *Report on the Social Cost of Greenhouse Gases: Estimates Incorporating Recent Scientific Advances* (2002).

42. See Env. Prot. Agency, *Report on the Social Cost of Greenhouse Gases: Estimates Incorporating Recent Scientific Advances* (2023), hereinafter Env. Prot. Agency (2023).

43. Id. at 14.

44. Id.

45. Id. at 15–16.

46. Id. at 16.

47. See generally Manuel Linsenmeier, Adil Mohommad, & Gregor Schwerhoff, *Global Benefits of the International Diffusion of Carbon Pricing Policies*, 13 Nature Climate Change 679 (2023); Robert C. Schmidt, Moritz Drupp, Frikk Nesje, & Hendrik Hoegen, *Testing the Free-Rider Hypothesis in Climate Policy* (Nov. 14, 2022), unpublished manuscript, on file at arXiv.org).

48. Env. Prot. Agency (2023), supra note 42, at 4.

49. Louisiana v. Biden, 585 F. Supp. 3d 840 (W.D. La. 2022).

50. Id. at 863, 863–865.

51. 472 F. Supp. 3d 573 (N.D. Cal. 2020).

52. 832 F. 3d 654 (7th Cir. 2016).

53. For details, including departures from the IAMs in certain respects, see Greenstone, Kopits, & Wolverton, supra note 21; Masur & Posner, supra note 9.

54. See William D. Nordhaus, *Economic Aspects of Global Warming in a Post-Copenhagen Environment*, 107 Proc. Nat'l Acad. Sci. U.S. 11721 (2010).

55. See David Anthoff & Richard S. J. Tol, *The Income Elasticity of the Impact of Climate Change*, in *Is the Environment a Luxury?* 34 (Silvia Tiezzi & Chiara Martini eds., 2014).

56. See Chris Hope, *The Social Cost of Co2 from the Page09 Model* (Econ. Discussion Paper No. 2011-39, 2011), https://papers.ssrn.com/sol3/papers.cfm?abstract_id=1973863.

57. See Greenstone, Kopits, & Wolverton, supra note 21.

58. Robert S. Pindyck, *Climate Change Policy: What Do the Models Tell Us?*, 51 J. Econ. Literature 860, 861 (2013); Robert S. Pindyck, *The Use and Misuse of Models for Climate Policy* (Nat'l Bureau of Econ. Rsch., Working Paper No. 21097, 2015).

59. See generally Env. Prot. Agency (2023), supra note 42.

60. Masur & Posner, supra note 9, at 1560.

61. Pindyck, *Climate Change Policy*, supra note 58, at 861.

62. Id. at 860.

63. Id. at 862.

64. Id. at 865.

65. Id. at 867.

66. Id.

67. Id. at 868.

68. Id. at 869.

69. Id. at 869–870.

70. See Carleton & Greenstone, supra note 6, offering an account of how to monetize the social cost of carbon in light of recent findings.

71. Robert S. Pindyck, *Climate Future: Averting and Adapting to Climate Change* (2022).

72. See Gayer & Viscusi, supra note 9, at 253.

73. It is natural to ask: Should benefits count too? This question contains an ambiguity. It could mean: Should the benefits to foreigners of emissions reduction be counted? If the analysis in the text is correct, that question is identical to this one: Should the costs of emissions matter to the analysis? (The answer to both questions is "yes.") But it could also mean: If (for example) polluting activity in the United States produces benefits to foreigners, should those benefits be counted? However that question is answered, it is independent of the question of whether the social *cost* of carbon should be measured in global or domestic terms. (A too-brief summary of what might be the outcome of an extended analysis: if regulation of polluting activity in the United States *removes benefits* to foreigners, it would be perfectly reasonable, consistent with the arguments presented here, to count those consequences as costs.)

74. See Jack Goldsmith, *Liberal Democracy and Cosmopolitan Duty*, 55 Stan. L. Rev. 1667, 1670–1671 (2003).

75. See id.; see also Simon Caney, *Justice Beyond Borders: A Global Political Theory* 3–16 (2005), exploring the issues from a philosophical perspective; Judith Lichtenberg, *National Boundaries and Moral Boundaries: A Cosmopolitan View*, in *Boundaries: National Autonomy and Its Limits* 79 (Peter G. Brown & Henry Shue eds., 1981), same. On cosmopolitanism and the social cost of carbon, see Masur & Posner, supra note 9, at 1593–1596.

76. See Carleton & Greenstone, supra note 6, at 4–5.

77. See Edna Ullmann-Margalit, *The Emergence of Norms* (1976), for an extended discussion.

78. See Elinor Ostrom, *Governing the Commons: The Evolution of Institutions for Collective Action* (1990); Robert Ellickson, *Order without Law: How Neighbors Settle Disputes* (1994).

79. See generally Linsenmeier, Mohommad, & Schwerhoff, supra note 48; Schmidt, Drupp, Nesje, & Hoegen, supra note 48; Trevor Houser, Kate Larsen, & Michael Greenstone, *Does the World Free Ride on US Pledges to Reduce Greenhouse Gas Emissions? Evidence from the Paris Climate Agreement* (Energy Pol'y Inst. at the Univ. of Chicago, Working Paper No. 2023-146, 2023), https://papers.ssrn.com/sol3/papers.cfm?abstract_id=4634315.

80. See Gayer & Viscusi, supra note 10; Jonathan Masur, *The Intractable Normative Complexities of Valuing Foreign Lives*, 2015 U. Ill. L. Rev. Slip Op. 12 (2015).

81. Id. at 14.

82. See Howard & Schwartz, supra note 8, at 223.

83. Motor Vehicle Mfrs. Ass'n of U.S., Inc. v. State Farm Mut. Auto. Ins. Co., 463 U.S. 29 (1983); Dep't of Homeland Sec. v. Regents of the Univ. of Cal., 140 S. Ct. 1891, 1905 (2020).

84. See Pindyck (2013), supra note 58, at 862, arguing that the SCC must include the possibility of catastrophic events, which IAMs as currently constructed do not anticipate.

85. See Stern & Stiglitz, supra note 9 at 8–9, discussing the analytical failures of IAMs.

86. See Jon Elster, *Sour Grapes* (1983).

## CHAPTER 2

1. See generally Erin Roberts & Saleemul Huq, *Coming Full Circle: The History of Loss and Damage under the UNFCCC*, 8 Int'l J. Glob. Warming 141 (2015); Edward A. Page & Clare Heyward, *Compensating for Climate Change Loss and Damage*, 65 Pol. Stud. 356 (2017); *Research Handbook on Climate Change Law and Loss & Damage* (Meinhard Doelle & Sara L. Seck eds., 2021); *Loss and Damage from Climate Change: Concepts, Methods and Policy Options* (Reinhard Mechler, Laurens M. Bouwer, Thomas Schinko, Swenja Surminski, & JoAnne Linnerooth-Bayer eds., 2019).

2. *Historic "Loss and Damage" Fund Adopted at COP27 Climate Summit*, Al Jazeera (Nov. 20, 2022), https://www.aljazeera.com/news/2022/11/20/historic-loss-and-damage-fund-adopted-at-cop27-climate-talks.

3. Id.

4. For more information on the political aspect of the loss and damage question, see E. Calliari, O. Serdeczny, & L. Vanhala, *Making Sense of the Politics in the Climate Change Loss & Damage Debate*, 64 Glob. Env't Change 1 (2020).

5. Valerie Volcovici, *Countries Deadlocked on "Loss and Damage" Fund as UN Climate Summit Nears*, Reuters (Oct. 23, 2023), https://www.reuters.com/sustainability/cop/countries-deadlocked-loss-damage-fund-un-climate-summit-nears-2023-10-23/.

6. Brad Plumer, *Carbon Dioxide Emissions Increased in 2022 as Crises Roiled Energy Markets*, N.Y. Times (Nov. 10, 2022). The United States and China account for 46 percent of global emissions.

7. Michon Scott, *Does It Matter How Much the United States Reduces Its Carbon Dioxide Emissions If China Doesn't Do the Same?*, Climate.gov (Aug. 30, 2023), https://www.climate.gov/news-features/climate-qa/does-it-matter-how-much-united-states-reduces-its-carbon-dioxide-emissions.

8. Laura Paddison & Annette Choi, *As Climate Chaos Accelerates, Which Countries Are Polluting the Most?*, CNN (Dec. 1, 2023), https://www.cnn.com/interactive/2023/12/us/countries-climate-change-emissions-cop28/. Based on 2022 data, "the average American is responsible for nearly twice as much climate pollution as the average person in China."

9. For valuable discussion, see William Nordhaus, *The Climate Casino* (2013); William Nordhaus, *The Spirit of Green* (2021).

10. See Nordhaus, *The Climate Casino*, supra note 9.

11. William D. Nordhaus & Joseph Boyer, *Warming the World: Economic Models of Global Warming* 152 (2000).

12. See Kelly Levin, David Waskow, & Rhys Gerholdt, *5 Big Findings from the IPCC's 2021 Climate Report*, World Res. Inst. (Aug. 9, 2021), https://www.wri.org/insights/ipcc-climate-report#:~:text=The%20IPCC%20Working%20Group%20I,on%20actions%20taken%20this%20decade; William Nordhaus, *The Challenge of Global Warming: Economic Models and Environmental Policy* 11 (2007), https://www.researchgate.net/publication/268396861_The_Challenge_of_Global_Warming_Economic_Models_and_Environmental_Policy.

13. For an estimate of the savings from a 0.03° reduction in warming, see Nordhaus and Boyer, supra note 11, at 156–167, suggesting $96 billion in worldwide benefits.

14. See Richard Stewart & Jonathan Wiener, *Reconstructing Climate Policy: Beyond Kyoto* 45–46 (2003).

15. This judgment is crude. If a high-emitting nation could reduce its emissions at relatively low cost, perhaps because of technological innovation, its burdens would of course be lower.

16. U.S. Energy Information Administration, "Table A10. World Carbon Dioxide Emissions by Region, Reference Case," *International Energy Outlook 2021* (Oct. 2021), https://www.eia.gov/outlooks/ieo/data/pdf/ref/A10_r.pdf.

17. See IPCC, *Climate Change 2007: The Physical Science Basis. Contribution of Working Group I to the Fourth Assessment Report of the Intergovernmental Panel on Climate Change, FAQ 10.3* (2007).

18. For various accounts, see Nicholas Stern, *The Economics of Climate Change: The Stern Review* (2007); IPCC, supra note 17.

19. Stern, supra note 18, at 139; Richard Tol, *Estimates of the Damage Costs of Climate Change*, 21 Env't & Res. Econ. 135 (2002).

20. Stern, supra note 18, at 139.

21. Id.

22. Tol, supra note 19, is in general accord. William Cline, *Climate Change*, in *Global Problems, Global Solutions* 13 (Bjorn Lomborg ed., 2004), and Frank Ackerman & Ian Finlayson, *The Economics of Inaction on Climate Change: A Sensitivity Analysis*, 6 Climate Pol'y 509 (2006), offer a picture of more serious monetized damage from climate change.

23. Id.

24. Yale Program on Climate Change Commc'n, *Climate Change in the Indian Mind, 2022* 5 (2022).

25. Enos Moyo, Leroy Gore Nhari, Perseverance Moyo, Grant Murewanhema, & Tafadzwa Dzinamarira, *Health Effects of Climate Change in Africa: A Call for an Improved Implementation of Prevention Measures*, 2 Eco-Env't & Health 74, 75 (2023).

26. Id.

27. See *The Economics of Climate Change*, Swiss Re Inst. (Apr. 22, 2021), https://www.swissre.com/institute/research/topics-and-risk-dialogues/climate-and-natural-catastrophe-risk/expertise-publication-economics-of-climate-change.html.

28. See id.; see also National Research Council, *Abrupt Climate Change: Inevitable Surprises* (2004); *Avoiding Dangerous Climate Change* (Hans Schellnhuber et al. eds., 2006).

29. See, e.g., Janet Currie & Firouz Gahvari, *Transfers in Cash and In-kind: Theory Meets the Data*, 46 J. Econ. Lit. 333 (2008); Shubhashis Gangopadhyay, Robert Lensink, & Bhupesh Yadav, *Cash or In-kind Transfers? Evidence from a Randomised Controlled Trial in Delhi, India*, 51 J. Development Studies, 660 (2015).

30. For various perspectives, see Steve Vanderheiden, *Globalizing Responsibility for Climate Change*, 25 Ethics & International Affairs 65 (2011); Dan Farber, *Basic Compensation for Victims of Climate Change*, 155 U. Pa. L. Rev. 1605 (2007); Kees van der Geest & Koko Warner, *Loss and Damage from Climate Change: Emerging Perspectives*, 8 Int. J. Global Warming 133 (2015).

31. Application of the Convention on the Prevention and Punishment of the Crime of Genocide (Bosnia and Herzegovina v. Serbia and Montenegro), Judgment, 2007 I.C.J. Rep. 43 (Feb. 26). The court ultimately denied Bosnia a remedy.

32. Thomas Schelling, *Intergenerational Discounting*, in *Discounting and Intergenerational Equity*, 99, 100 (Paul Portney & John P. Weyant eds., 1999).

33. See, e.g., H. D. Lewis, *Collective Responsibility (A Critique) in Collective Responsibility: Five Decades of Debate*, in *Theoretical & Applied Ethics* 17–34 (Larry May & Stacey Hoffman eds., 1991). In recent years, some philosophers have challenged traditional criticisms of collective responsibility, but these philosophers tend to ground collective responsibility in individual failures to act when action was possible and likely to be effective, and the person in question knew or should have known that they could have prevented the harm. See, e.g., Larry May, *Sharing Responsibility* (1992). Cf. Brent Fisse & John Braithwaite, *Corporations, Crime and Accountability* 50 (1993); Christopher Kutz, *Complicity: Ethics and Law for a Collective Age* 166–253 (2000); David Copp, *Responsibility for Collective Inaction*, 22 J. Soc. Phil. 71, 71 (1991).

34. Stephen Kershnar, *The Inheritance-Based Claim to Reparations*, 8 Legal Theory 243, 251–257 (2002). These arguments are often analogized to unjust enrichment arguments. See Eric A. Posner & Adrian Vermeule, *Reparations for Slavery and Other Historical Injustices*, 103 Colum. L. Rev. 689 (2003).

35. See, e.g., David Herring, *Are There Positive Benefits From Global Warming?*, Climate.gov (Oct. 29, 2020), https://www.climate.gov/news-features/climate-qa/are-there-positive-benefits-global-warming.

36. See R. A. Pielke et al., *Hurricanes and Global Warming*, 86 Bull. of the Am. Meteorological Soc'y 1571 (2005) (discussing the dubious connection between increased hurricane intensity and climate change).

37. See Michael Saks & Peter Blanck, *Justice Improved: The Unrecognized Benefits of Aggregation and Sampling in the Trial of Mass Torts*, 44 Stan. L. Rev. 815 (1992).

38. Compare Jules Coleman, *Tort Law and the Demands of Corrective Justice*, 67 Indiana L.J. 349 (1992), arguing that corrective justice requires a remedy even when the infringing conduct was innocent, with Ernest Weinrib, *Corrective Justice*, 77 Iowa L. Rev. 403 (1992), taking the contrary view. For a very helpful discussion, see Stephen R. Perry, *Loss, Agency, and Responsibility for Outcomes: Three Conceptions of Corrective Justice*, in *Tort Theory* 24 (Ken Cooper-Stephenson & Elaine Gibson eds., 1993).

39. See Richard A. Posner, *Economic Analysis of Law* (1973).

40. For simplicity, we will rely on the legal view. However, the legal standard does not, strictly speaking, require culpability. See A. P. Simester, *Can Negligence Be Culpable?*, in *Oxford Essays in Jurisprudence* 85, 87 (Jeremy Horder ed., 2000).

41. One commentator suggests 1990 as a date for when emitting activities could have become negligent. See Jiahua Pan, *Common but Differentiated Commitments: A Practical Approach to Engaging Large Developing Emitters Under L20* 3–7 (2004).

42. See National Development and Reform Commission, People's Republic of China, *China's National Climate Change Programme* 58 (June 2007).

43. See Ying Chen & Jiahua Pan, *Equity Concerns over Climate Change Mitigation* (Chinese Academy of Social Sciences, Global Change and Economic Development Program Working Paper No. 002).

44. On the general idea, formalized at the 1992 United Nations Conference on Environment and Development in Rio de Janeiro, see Lavanya Rajamani, *The Principle of Common but Differentiated Responsibility and the Balance of Commitments under the Climate Regime*, 9 Rev. Eur. Comp. & Int'l Envtl. L. 120 (2000); Philippos C. Spyropoulos & Theodore P. Fortsakis, *The Common but Differentiated Responsibility Principle in Multilateral Environmental Agreements: Regulatory and Policy Aspects* (2009). On the link, see National Development and Reform Commission, supra note 42, at 58.

45. See Pan, supra note 41, at 3–4.

46. See Emanuele Campiglio, *Who Should Pay for Climate? The Effect of Burden-Sharing Mechanisms on Abatement Policies and Technological Transfers* (Grantham Research Institute on Climate Change and the Environment Working Paper No. 96, 2012); Elisa Calliari, Olivia Serdeczny, & Lisa Vanhala, *Making Sense of the Politics in the Climate Change Loss and Damage Debate*, 64 Global Envtl. Change, 6–8 (2020); Daniel A. Farber, *Adapting to Climate Change: Who Should Pay*, 23 J. Land Use & Envtl. L. 1, 25–30 (2007); *Should Rich Countries Pay for Climate Damage in Poor Ones?*, The Economist (Nov. 24, 2022), https://www.economist.com/international/2022/11/20/a-new-un-fund-for-loss-and-damage-emerges-from-cop27.

47. See Campiglio, supra note 46; Calliari, Serdeczny, & Vanhala, supra note 46.

48. Id.

## CHAPTER 3

1. Off. of Mgmt. & Budget, Exec. Off. of the President, Circular A-4, To the Heads of Executive Agencies and Establishments: Regulatory Analysis 15 (2003).

2. Nicholas Stern, *The Economics of Climate Change* (2007); William Nordhaus, *A Question of Balance* (2008).

3. See Nordhaus, supra note 2; *The Marginal Impact of $CO_2$ from PAGE2002: An Integrated Assessment Model Incorporating the IPCC's Five Reasons for Concern*, 6 Integrated Assessment J. 19 (2006).

4. IPCC, *Climate Change 2007—Impacts, Adaptation and Vulnerability* 823 (2007).

5. Frank Ramsey, *A Mathematical Theory of Savings*, 38 Econ. J. 543 (1928).

6. Roy Harrod, *Towards a Dynamic Economics* (1948).

7. Tjalling C. Koopmans, *On the Concept of Optimal Economic Growth*, 28 Pontificae Academiae Scientiarum Scripta Varia 225 (1965).

8. One especially good collection is *Discounting and Intergenerational Equity* (Paul Portney & John P. Weyant eds., 1999).

9. Classic references include Kenneth J. Arrow & Robert C. Lind, *Uncertainty and the Evaluation of Public Investment Decisions*, 60 Am. Econ. Rev. 364 (1970); David Bradford, *Constraints on Government Investment Opportunities and the Choice of the Discount Rate*, 65 Am. Econ. Rev. 887 (1975); S. A. Marglin, *The Opportunity Costs of Public Investment*, 77 Q. J. of Econ. 274 (1963); Robert C. Lind, *A Primer on the Major Issues Relating to the Discount Rate for Evaluating National Energy Options*, in *Discounting for Time and Risk in Energy Policy* 21 (Robert C. Lind et al. ed., 1982).

10. M. L. Weitzman, *Why the Far-Distant Future Should Be Discounted at Its Lowest Possible Rate*, 36 J. of Env't Econ. & Mgmt. 201 (1998).

11. See the discussion of the just savings principle in John Rawls, *A Theory of Justice* (1971).

12. This simplification is simply astonishing. Stern, for example, severely criticizes the positivists for requiring all kinds of specialized assumptions for the private rate of return to equal the social rate of return, but then imposes specialized functional forms (Stern, supra note 2). Although the use of this functional form has a long history in public economics, it remains a specialized assumption. See Anthony B. Atkinson, *Measurement of Inequality*, 2 J. of Econ. Theory 244 (1970).

13. Stern, supra note 2.

14. If it turns out that the project is desirable—that current projects have lower rates of return—then there is an issue about how to divide the surplus. Any division makes both generations better off.

15. It is necessary to use $95 rather than $100 because the newly discovered environmental harm means that we (all generations together) are not as well off as we thought. It is as if we lost money. It is likely that all generations will need to share in this loss. This number is only illustrative and entails no position on the damages from climate change.

16. This is a simplification of the idea of distribution-neutral investment choice, discussed in detail in Louis Kaplow, *Discounting Dollars, Discounting Lives: Intergenerational Distributive Justice and Efficiency*, 74 U. Chi. L. Rev. 79 (2007); Dexter Samida & David A. Weisbach, *Paretian Intergenerational Discounting*, 74 U. Chi. L. Rev. 145 (2007).

17. An alternative, slightly more controversial way to make this point is that the ethicists observe that the private rate of return is not equal to the social rate of return and suggest that the government can fill this gap. For example, if the private market rejects a project because the rate of return is only, say, 5 percent when it

demands a 5.5 percent return, then the government should engage in the project if the social rate of return is lower, such as the 1.4 percent used by Stern. Given large differences in the private rate of return and the social rate of return, the government would be engaging in a vastly greater number of projects than any democratic government currently does. There are likely to be good reasons for restricting the scope of government projects, however. Therefore, the ethicists' arguments for a very low social discount rate are incomplete. Recommendations about government projects using a low social discount rate need to be combined with these reasons for restricting government projects. The models run by the ethicists and the resulting recommendations, however, never include these exogenous restrictions.

18. A more subtle and less powerful objection is that even if overall savings rates stay the same, interest rates may change with a change in projects. For example, if we keep our legacy to the future at $100 but change the mix of projects that make up this $100, then market rates of return may change. This concern seems second order, and interest rates could go up as well as down. See Kaplow, supra note 16, for a discussion.

19. Robert J. Barro & Xavier Sala-i-Martin, *Economic Growth* (1995).

20. For a review, see Kent Smetters, *Ricardian Equivalence: Long-Run Leviathan*, 73 J. of Pub. Econ. 395 (1999).

21. Lind, *A Primer*, supra note 9; Robert C. Lind, *Analysis for Intergenerational Discounting*, in *Discounting and Intergenerational Equity*, at 173, 176.

22. Lind, *Analysis for Intergenerational Discounting*, supra note 21.

23. Richard L. Revesz, *Environmental Regulation, Cost-Benefit Analysis, and the Discounting of Human Lives*, 99 Colum. L. Rev. 941 (1999).

24. See, e.g., W. Kip Viscusi, *Fatal Tradeoffs* (1994).

25. Derek Parfit, *Reasons and Persons* (1984).

## CHAPTER 4

1. Office of Management and Budget, Circular A-4 (2023), available at https://www.whitehouse.gov/wp-content/uploads/2023/11/CircularA-4.pdf.

2. See Env. Prot. Agency, *Report on the Social Cost of Greenhouse Gases: Estimates Incorporating Recent Scientific Advances* (2023), available at https://www.epa.gov/system/files/documents/2023-12/epa_scghg_2023_report_final.pdf. For discussion, see Dylan Matthews, The Tricky Business of Putting a Dollar Value on a Human Life (2022), available at https://www.vox.com/future-perfect/23449849/social-cost-carbon-value-statistical-life-epa.

3. The literature appears to begin with Richard Thaler & Sherwin Rosen, *The Value of Saving a Life: Evidence from the Labor Market*, in *Household Production and Consumption* 265 (Nestor E. Terleckyj ed., 1976).

4. See, e.g., W. Kip Viscusi & Joseph Aldy, *The Value of a Statistical Life*, 27 J. Risk & Uncertainty 5 (2003); W. Kip Viscusi, *The Heterogeneity of the Value of a Statistical Life: Introduction and Overview*, 40 J. Risk & Uncertainty 1 (2010), noting the median value of $7 million to $8 million.

5. For examples in an important and unusually interesting setting, see Sean Hannon Williams, *Statistical Children*, 30 Yale J. Regul. 63 (2013).

6. I discuss some of them in Cass R. Sunstein, *The Cost-Benefit Revolution* (2018).

7. See Viscusi, supra note 3; Kyle Greenberg et al., *The Heterogeneous Value of a Statistical Life: Evidence from U.S. Army Reenlistment Decisions* (Nat'l Bureau of Econ. Rsch., Working Paper No. 29104, 2021).

8. See Notice of Availability: Proposed Draft Guidance for Estimating Value per Statistical Life, 88 Fed. Reg. 17,826 (Mar. 24, 2023), setting VSL at $11.6 million; Agamoni Majumder & S. Madheswaran, *Value of Statistical Life in India: A Hedonic Wage Approach* (Inst. for Soc. & Econ. Change, Working Paper No. 407, 2018), setting VSL at $0.64 million.

9. See Daniel Hemel, *Regulation and Redistribution with Lives in the Balance*, 89 U. Chi. L. Rev. 649 (2022).

10. See Louis Kaplow & Steven Shavell, *Why the Legal System Is Less Efficient than the Income Tax in Redistributing Income*, 23 J. Legal Stud. 667 (1994). On some of the complexities, see Zachary Liscow, *Redistribution for Realists*, 107 Iowa L. Rev. 495 (2022); Richard L. Revesz, *Regulation and Distribution*, 93 N.Y.U. L. Rev. 1489 (2018); Matthew D. Adler, *Benefit-Cost Analysis and Distributional Weights: An Overview*, 10 Rev. Env. Econ. & Pol'y 264 (2016).

11. See Zachary Liscow, *Redistribution for Realists*, 107 Iowa L. Rev. 495 (2022).

12. For relevant discussion, see Daniel Hemel, *Regulation and Redistribution with Lives in the Balance*, 89 U. Chi. L. Rev. 649 (2022). See also David Harrison Jr., *Who Pays for Clean Air: The Cost and Benefit Distribution of Federal Automobile Emission Standards* (1975), finding that the cost of automobile emission controls as a proportion of income is larger for lower-income households than for higher-income households, while distribution of benefits by income is less clear; Robert Dorfman, *Incidence of the Benefits and Costs of Environmental Programs*, 67 Am. Econ. Rev. 333 (1977), estimating the benefits of pollution control based on self-reported willingness to pay and finding that pollution control imposes net costs on lower-income households and yields net benefits for higher-income households; Leonard P. Gianessi, Henry M. Peskin, & Edward Wolff, *The Distributional Effects of Uniform Air Pollution Policy in the United States*, 93 Q.J. Econ. 281 (1979), finding that industrial-air-pollution controls impose net costs on higher-income households and generate net benefits for lower-income households while automobile emissions controls impose net costs on all income groups; Matthew E. Kahn, *The Beneficiaries of Clean Air Act Regulation*,

24 Regul. 34 (2001) ("It appears that regulation under the Clean Air Act has helped, and not economically harmed, the 'have nots.'").

13. See Kelly McGee, *A Place Worth Protecting: Rethinking Cost-Benefit Analysis Under FEMA's Flood-Mitigation Programs*, 88 U. Chi. L. Rev. 1925 (2021).

14. See Cass R. Sunstein, *Willingness to Pay versus Welfare*, 1 Harv. L. & Pol'y Rev. 303 (2007).

15. See Daniel Hemel, *Regulation and Redistribution with Lives in the Balance*, 89 U. Chi. L. Rev. 649 (2022).

16. I am assuming that there is no other loss for construction workers from the regulation, such as loss of employment, wage reductions, or reductions in benefits provided by employers.

17. I am bracketing possible incentive effects: a program that helps the poor, and hurts the rich, might decrease the incentive to be rich rather than poor. But especially insofar as we are speaking of funding programs for (for example) climate-related risks, any such incentive effects are likely to be modest.

18. See Matthew D. Adler & Ole F. Norheim, *Prioritarianism in Practice* (2022).

## CHAPTER 5

1. Patrick W. Baylis & Judson Boomhower, *Mandated vs. Voluntary Adaptation to Natural Disasters: The Case of U.S. Wildfires* 1 (Nat'l Bureau of Econ. Rsch., Working Paper No. 29621, 2021).

2. Id. at 2.

3. Id. at 2, 7.

4. Id. at 2.

5. See Ellen Vaughan & Jim Turner, Env't & Energy Study Inst., *The Value and Impact of Building Codes* 1–3 (2013); *5 Reasons Building Codes Should Matter to You*, Fed. Emergency Mgmt. Agency (Sept. 29, 2021), https://www.fema.gov/blog/5-reasons-building-codes-should-matter-you; Fed. Emergency Mgmt. Agency, *Building Codes Save: A Nationwide Study*, at ES-1 to ES-3 (2020); Ins. Inst. for Bus. & Home Safety, *Building Code Resources: The Benefits of Statewide Building Costs* (2019); Env't Prot. Agency, *Smart Growth Fixes for Climate Adaptation and Resilience* 1–4 (2017).

6. See United Nations, *Towards COP27: Compendium of Climate-Related Initiatives* 7 (2022).

7. See Jane E. Leggett, Cong. Rsch. Serv., IF11827, *Climate Change: Defining Adaptation and Resilience, with Implications for Policy* (2011).

8. See UN Sys. Task Team on Post-2015 UN Dev. Agenda, *Disaster Risk and Resilience* 3 n.1 (2012).

9. See IPCC, *Annex II: Glossary* 1758 (2018).

10. See *Building Resilient Infrastructure and Communities*, Fed. Emergency Mgmt. Agency, https://www.fema.gov/grants/mitigation/building-resilient-infrastructure-communities.

11. See *National Risk Index for Natural Hazards*, Fed. Emergency Mgmt. Agency, https://www.fema.gov/flood-maps/products-tools/national-risk-index; *Nature-Based Solutions*, Fed. Emergency Mgmt. Agency, https://www.fema.gov/emergency-managers/risk-management/nature-based-solutions; Flood Factor, https://floodfactor.com/; Henry Fountain, *This Vast Wildlife Lab Is Helping Foresters Prepare for a Hotter Planet*, N.Y. Times (Jan. 5, 2022), https://www.nytimes.com/2022/01/05/climate/fire-forest-management-bootleg-oregon.html.

12. See Zhongchen Hu, *Salience and Households' Flood Insurance Decisions* 1–3 (Feb. 22, 2021), unpublished manuscript, on file with author.

13. See generally Cass R. Sunstein, *Sludge: What Stops Us from Getting Things Done and What to Do about It* (2021), exploring administrative burdens and their adverse effects.

14. See Lynn Conell-Price, *Encouraging Resiliency with Auto-Enrollment in Supplemental Flood Insurance Coverage*, Wharton Risk Mgmt. & Decision Processes Ctr. (June 17, 2021), https://esg.wharton.upenn.edu/news/encouraging-resiliency-with-auto-enrollment-in-supplemental-flood-insurance-coverage/.

15. Shlomo Benartzi, John Beshears, Katherine L. Milkman, Cass R. Sunstein, Richard H. Thaler, Maya Shankar, Will Tucker-Ray, William J. Congdon, & Steven Galing, *Should Governments Invest More in Nudging?*, 28 Psych. Sci. 1041, 1044–1052 (2017).

16. Stephanie Mertens, Mario Herberz, Ulf J. J. Hahnel, & Tobias Brosch, *The Effectiveness of Nudging: A Meta-Analysis of Choice Architecture Interventions across Behavioral Domains*, 119 Proc. Nat'l. Acad. Sci. 1 (2022).

17. Cass R. Sunstein, *The Distributional Effects of Nudges*, 6 Nature Hum. Behav. 9 (2022).

18. See generally Nicholas Stern & Joseph E. Stiglitz, *The Social Cost of Carbon, Risk, Distribution, Market Failures: An Alternative Approach* 1 (Nat'l Bureau of Econ. Rsch., Working Paper No. 28472, 2021), arguing for a ceiling on emissions and a social cost of carbon that is a product of that ceiling.

19. Id. at 59.

20. See id. at 2; see, e.g., William D. Nordhaus, *Projections and Uncertainties about Climate Change in an Era of Minimal Climate Policies*, 10 Am. Econ. J.: Econ. Pol'y 333, 333–360 (2018).

21. See Robert S. Pindyck, *Climate Change Policy: What Do the Models Tell Us?*, 51 J. Econ. Literature 860, 870 (2013); Robert S. Pindyck, *What We Know and Don't Know*

*about Climate Change, and Implications for Policy*, 2 Env't & Energy Pol'y & Econ. 4, 17–26 (2021).

22. See Martin L. Weitzman, *Fat-Tailed Uncertainty in the Economics of Catastrophic Climate Change*, 5 Rev. Env't Econ. & Pol'y 275, 279–283 (2011).

23. See, e.g., Robert S. Pindyck, *Climate Futures* 8 (2022).

24. Id.

25. I discuss some of the issues here in Cass R. Sunstein, *Averting Catastrophe: Decision Theory for COVID-19, Climate Change, and Potential Disasters of All Kinds* (2021).

26. See Baylis & Boomhower, supra note 1, at 32.

27. See generally Matthew D. Adler, *Theory of Prioritarianism*, in *Prioritarianism in Practice* (Matthew D. Adler & Ole F. Norheim eds., 2022), outlining the theory of prioritarianism as a branch of welfare consequentialism; Matthew D. Adler & Nils Holtug, *Prioritarianism: A Response to Critics*, 18 Pol. Phil. & Econ. 101 (2019), defending prioritarianism against objections.

28. This paragraph borrows heavily from a discussion in Cass R. Sunstein, *Arbitrariness Review and Climate Change*, 170 U. Pa. L. Rev 991 (2022).

29. John Rawls, *A Theory of Justice* 83 (1971).

30. John Maynard Keynes, *The General Theory of Employment, Interest and Money* 113–114 (1936).

31. See Frank H. Knight, *Risk, Uncertainty and Profit* (1921).

32. Pindyck, supra note 23, at 181.

33. Some valuable information can be found in Fed. Emergency Mgmt. Agency, *FEMA Resources for Climate Resilience* (2021).

34. In the context of flooding, see Flood Factor, https://floodfactor.com/. In the context of a large number of risks, see *National Risk Index for Natural Hazards*, Fed. Emergency Mgmt. Agency, https://www.fema.gov/flood-maps/products-tools/national-risk-index.

35. See *Building Resilient Infrastructure and Communities*, Fed. Emergency Mgmt. Agency (Oct. 20, 2023), https://www.fema.gov/grants/mitigation/building-resilient-infrastructure-communities.

## CHAPTER 6

1. Hunt Allcott & Judd Kessler, *The Welfare Effects of Nudges: A Case Study of Energy Use Social Comparisons*, 11 Applied Econ. 236 (2019), available at https://www.aeaweb.org/articles?id=10.1257/app.20170328.

2. Hunt Allcott, Daniel Cohen, William Morrison, & Dmitry Taubinsky, *When Do "Nudges" Increase Welfare?* (Nat'l Bureau of Econ. Rsch., Working Paper No. 30740, 2022).

3. Linda Thunström, *Welfare Effects of Nudges: The Emotional Tax of Calorie Menu Labeling*, 14 Judgment & Decision Making 11 (2019).

4. See Allcott, Cohen, Morrison, & Taubinsky, supra note 2.

5. The term is not in general use, but something like it can be found in various places, with variations. See Michael Yeomans, Anuj Shah, Sendhil Mullainathan, & Jon Kleinberg, *Making Sense of Recommendations*, 32 J. Behav. Decision Making 403 (2019); Guy Champniss, *The Rise of the Choice Engine*, Enervee (Mar. 6, 2018), https://www.enervee.com/blog/the-rise-of-the-choice-engine. Compare the following, which is regrettably complicated: *Buying a Refrigerator Guide*, Whirlpool, https://www.whirlpool.com/blog/kitchen/buying-guide-refrigerator.html.

6. See Vidya S. Athota, Zahid Hasan, Daicy Vaz, Sop Sop Maturin Désiré, & Vijay Pereira, *Can Artificial Intelligence (AI) Manage Behavioural Biases among Financial Planners?*, 31 J. Glob. Info. Mgmt. 1 (2023). For a disturbing set of findings, see Yang Chen, Samuel Kirshner, Anton Ovchinnikov, Meena Andiappan, & Tracy Jenkin, *A Manager and an AI Walk into a Bar: Does ChatGPT Make Biased Decisions Like We Do?* (2023), available at https://papers.ssrn.com/sol3/papers.cfm?abstract_id=4380365.

7. See Jamie Luguri & Lior Strahilevitz, *Shining a Light on Dark Patterns*, 13 J. Legal Analysis 43 (2021).

8. See, e.g., Joachim Schleich, Xavier Gassmann, Thomas Meissner, & Corinne Faure, *A Large-Scale Test of the Effects of Time Discounting*, 80 Energy Economics 377 (2019); Madeline Werthschulte & Andreas Löschel, *On the Role of Present Bias and Biased Price Beliefs in Household Energy Consumption*, 109 J Env. Ec. and Management (2021); Theresa Kuchler & Michaela Pagel, *Sticking to Your Plan: The Role of Present Bias for Credit Card Paydown* (2018), available at https://www.nber.org/system/files/working_papers/w24881/w24881.pdf; Ted O'Donoghue & Matthew Rabin, *Present Bias: Lessons Learned and to Be Learned*, 105 Am Econ Rev 273 (2015); Jess Benhabib, Alberto Bisin, & Andrew Schotter, *Present Bias, Quasi-Hyperbolic Discounting, and Fixed Costs*, 69 Games and Economic Behavior 205 (2010); Yang Wang & Frank Sloan, *Present Bias and Health*, J. Risk and Uncertainty 177 (2018). Importantly, Wang and Sloan find strong evidence of present bias in connection with health-related decisions.

9. It might. Chen, Kirshner, Ovchinnikov, Andiappan, & Jenkin, supra note 6.

10. See Ian Ayres & Quinn Curtis, *Retirement Guardrails: How Proactive Fiduciaries Can Improve Plan Outcomes* (2023).

11. See Cass R. Sunstein, *Why Nudge?* (2013).

12. See Kenneth T. Gillingham, Sébastien Houde, & Arthur A. van Benthem, *Consumer Myopia in Vehicle Purchases: Evidence from a Natural Experiment*, 13 Am. Econ. J.: Econ. Pol'y 207 (2021).

13. Id. at 207.

14. See Valerie J. Karplus, Sergey Paltsev, Mustafa Babiker, & John M. Reilly, *Should a Vehicle Fuel Economy Standard Be Combined with an Economy-Wide Greenhouse Gas Emissions Constraint? Implications for Energy and Climate Policy in the United States*, 36 Energy Econ. 322 (2013); Christopher R. Knittel, *Diary of a Wimpy Carbon Tax* (MIT Ctr. for Energy & Envtl. Policy Research, Working Paper No. 13, 2019), https://ceepr.mit.edu/workingpaper/diary-of-a-wimpy-carbon-tax-carbon-taxes-as-federal-climate-policy/; Lucas W. Davis and Christopher R. Knittel, *Are Fuel Economy Standards Regressive?* (Nat'l Bureau of Econ. Research, Working Paper No. 22925, 2016), https://www.nber.org/papers/w22925.

15. Karplus, Paltsev, Babiker, & Reilly, supra note 14, at 322.

16. See Richard P. Larrick & Jack B. Soll, *The MPG Illusion*, 320 Sci. 1593, 1593 (2008).

17. Nat'l. Highway Traffic Safety Admin., *Final Regulatory Impact Analysis: Corporate Average Fuel Economy for MY 2017–MY 2025* (2012), at 49–50.

18. See Antonio M. Bento, Mark R. Jacobsen, Christopher R. Knittel, & Arthur A. van Benthem, *Estimating the Costs and Benefits of Fuel Economy Standards*, 1 Envtl. and Energy Pol. and the Econ. 129 (2020).

19. See Hunt Allcott & Christopher Knittel, *Are Consumers Poorly Informed about Fuel Economy?*, 11 Am. Econ. J. Econ. Pol. 1 (2019); James M. Sallee, Sarah E. West, & Wei Fan, *Do Consumers Recognize the Value of Fuel Economy? Evidence from Used Car Prices and Gasoline Price Fluctuations*, 135 J. Pub. Econ. 61 (2016); Meghan R. Busse, Christopher R. Knittel, & Florian Zettelmeyer, *Are Consumers Myopic? Evidence from New and Used Car Purchases*, 103 Am. Econ. Rev. 220 (2013).

20. See Bento, Jacobsen, Knittel, & van Benthem, supra note 18.

21. See Revised 2023 and Later Model Year Light-Duty Vehicle Greenhouse Gas Emissions Standards, 86 Fed. Reg. 74,434 (Dec. 30, 2021). See in particular id. at 74,500 et seq., discussing the evaluation of consumer impacts.

22. See Ted Gayer & W. Kip Viscusi, *Overriding Consumer Preferences with Energy Regulations*, 43 J. Regul. Econ. 248, 254, 257 (2013).

23. See Sarah Conly, *Against Autonomy: Justifying Coercive Paternalism* (2012); Ryan Bubb & Richard H. Pildes, *How Behavioral Economics Trims Its Sails and Why*, 127 Harv. L. Rev. 1593 (2014).

24. See generally Hunt Allcott and Cass R. Sunstein, *Regulating Internalities*, 34 J. Pol'y. Analysis & Mgmt. 698 (2015).

25. See Karplus, Paltsev, Babiker, & Reilly, supra note 14, at 322.

26. See Cass R. Sunstein, *Rear Visibility and Some Unresolved Problems for Economic Analysis*, 10 J. Benefit-Cost Analysis 317 (2019).

27. For suggestive evidence, see Richard G. Newell & Juha V. Siikamäki, *Individual Time Preferences and Energy Efficiency* (Nat'l Bureau of Econ. Rsch., Working Paper No. 20969, 2015). Note that the miles-per-gallon measure is hardly hidden, and there is nothing quite as salient for energy efficiency.

28. See Light-Duty Vehicle Greenhouse Gas Emission Standards and Corporate Average Fuel Economy Standards; Final Rule, Part II, 75 Fed. Reg. 25,324, 25,510–511 (May 7, 2010) (to be codified at 40 C.F.R. pts. 85, 86, and 600; 49 C.F.R. pts. 531, 533, 536, et al.), http://www.gpo.gov/fdsys/pkg/FR-2010-05-07/pdf/2010-8159.pdf.

29. See supra note 21, at 74,500.

30. Id. at 74,501.

31. Id.

32. For valuable, inconclusive discussions, see generally Hunt Allcott, *Paternalism and Energy Efficiency: An Overview*, 8 Ann. Rev. Econ. 145 (2016); Hunt Allcott & Michael Greenstone, *Is There an Energy Efficiency Gap?*, 26 J. Econ. Persp. 3, 5 (2012); Allcott & Knittel, supra note 19; Sallee, West, & Fan, supra note 19; Busse, Knittel, & Zettelmeyer, supra note 19.

33. See Sallee, West, & Fan, supra note 19, at 61; Busse, Knittel, & Zettelmeyer, supra note 19, at 220.

34. See Allcott and Knittel, supra note 19, at 33–34.

35. See Denvil Duncan, Arthur Lin Ku, Alyssa Julian, Sanya Carley, Saba Siddiki, Nikolaos Zirogiannis, & John D. Graham, *Most Consumers Don't Buy Hybrids: Is Rational Choice a Sufficient Explanation?*, 10 J. Benefit-Cost Analysis 1, 1 (2019).

36. See id. at 30.

37. See John D. Graham & Jonathan B. Wiener, *Co-Benefits, Countervailing Risks, and Cost-Benefit Analysis*, in *Human and Ecological Risk Assessment: Theory and Practice* 1167, 1172 (Dennis J. Paustenbach ed., 3rd ed., 2024).

38. See Hunt Allcott, Benjamin B. Lockwood, & Dmitry Taubinsky, *Regressive Sin Taxes, with an Application to the Optimal Soda Tax*, 134 Q J. Econ. 1557 (2019).

39. Oren Bar-Gill, Cass R. Sunstein, & Inbal Talgam-Cohen, *Algorithmic Harm in Consumer Markets* (2022), available at https://papers.ssrn.com/sol3/papers.cfm?abstract_id=4321763.

40. On the general problem, see Cass R. Sunstein, *Manipulation as Theft*, 29 J. of European Public Policy 1959 (2022).

41. See Glenn Whitman & Mario Rizzo, *Escaping Paternalism* (2020).

42. See Chen, Kirshner, Ovchinnikov, Andiappan, & Jenkin, supra note 6.

43. See Ayres and Curtis, supra note 10.

44. See Saurabh Bhargava, George Loewenstein, & Justin Sydnor, *Choose to Lose: Health Plan Choices from a Menu with Dominated Option*, 132 Q. J. Econ. 1319 (2017).

45. See George A. Akerlof & Robert J. Shiller, *Phishing for Phools* (2015).

## EPILOGUE

1. There is some support here for a controversial claim in Derek Parfit, *On What Matters*, vol. 1 (2007).

# INDEX

Page numbers in italics refer to figures.